TURFGRASS PESTS

Compiled by
A. D. Ali
Extension Entomologist, University of California, Riverside
and Clyde L. Elmore
Extension Weed Scientist, University of California, Davis

COOPERATIVE EXTENSION
UNIVERSITY OF CALIFORNIA
DIVISION OF AGRICULTURE
AND NATURAL RESOURCES

For information about ordering this publication
or about ordering companion Leaflet 2209,
Guide to Turfgrass Pest Control, write to:

Publications
Division of Agriculture and Natural Resources
University of California
6701 San Pablo Avenue
Oakland, California 94608-1239

or telephone (415) 642-2431

Publication 4053

ISBN 0-931876-86-9
Library of Congress Catalog Card Number: 88-83113

© 1989 by the Regents of the University of California
Division of Agriculture and Natural Resources

All Rights reserved.

No part of this publication may be reproduced, stored in a retrieval system, or transmitted, in any form or by any means, electronic, mechanical, photocopying, recording, or otherwise, without the written permission of the publisher and the authors.

Printed in the United States of America.

Production:
Jim Coats, Senior Editor;
Franz Baumhackl, Senior Artist.

Photography:
Jack Kelly Clark,
and the authors.

To simplify information, trade names of products have been used. No endorsement of named products is intended, nor is criticism implied of similar products that are not mentioned.

Contents

Preface v

1. *Safe and Effective Use of Pesticide Chemicals*
   J. BLAIR BAILEY 1

2. *Measurements, Calculations, and Sprayer Preparation*
   JOHN VAN DAM AND DAVID W. CUDNEY 7

3. *Weed Control in Large Turf Areas*
   CLYDE L. ELMORE, W. B. MCHENRY, AND DAVID W. CUDNEY 17

4. *Insect and Related Pests of Turfgrass*
   A. D. ALI 49

5. *Nematode Diseases*
   JOHN D. RADEWALD AND BECKY B. WESTERDAHL 69

6. *Fungal Diseases*
   ARTHUR H. MCCAIN, ROBERT M. ENDO, AND HOWARD D. OHR 75

7. *Rodents and Related Vertebrate Pests*
   TERRELL SALMON 93

   Glossary 117

Preface

In this second revised edition of Turfgrass Pests, University of California farm advisors, specialists, and faculty again join forces to help professionals and amateurs deal with the pests that invade turfgrass. Though greater in scope and more up to date than the 1980 edition, the publication still emphasizes that "a chain is no stronger than its weakest link, and pest control practices will be no more effective than the attention paid to the least detail."

Photographs, many in color, again help the user identify insect, disease, weed, and vertebrate pests of turfgrass; the calendar of pest activity and treatment and the other tables provide quick and useful references. Considerable information is provided on the best selection of pesticides, but the most important information may be in Chapter 1, Safe and Effective Use of Pesticide Chemicals.

Most chapters have been revised and improved. A metric (SI) conversion table is included in Chapter 2 for your convenience.

This publication does not give specific chemical recommendations. Such recommendations are made in the companion *Guide to Turfgrass Pest Control*, Leaflet 2209, also published by the University of California's Division of Agriculture and Natural Resources. The leaflet is revised and reissued as needed. Copies may be obtained from your county farm advisor's office or by mail from the address printed on the back of the title page of this book.

1

Safe and Effective Use of Pesticide Chemicals

J. BLAIR BAILEY, EXTENSION ENTOMOLOGIST,
UNIVERSITY OF CALIFORNIA, RIVERSIDE

Pesticides are not only useful—in many cases they are absolutely necessary to the production of pest-free turfgrass. We can benefit from their proper use, but if pesticides are misused through carelessness or ignorance, the results can be expensive and dangerous. Before discussing the problems that can result from misuse, let's first get a clear idea of what pesticides are and, in general terms, how they work.

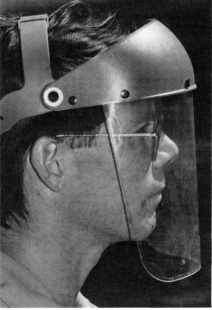

Wear protective gear—when specified on the pesticide label.

1

WHAT IS A PESTICIDE?

The term "pesticide"—defined legally as an "economic poison"—refers to any chemical or mixture of chemicals intended to control pests. For example:

PESTICIDES USED	PESTS CONTROLLED
insecticides	insects
fungicides	fungi and bacteria
herbicides	weeds
nematicides	nematodes
rodenticides	rodents

In general, pesticides control both plant and animal pests by poisoning them, so all pesticides are in some degree toxic to plants, animals, or both. Since we are animals, pesticides can poison us, too. If used incorrectly, some pesticides can kill desirable plants such as ground cover and shrubs and desirable animals such as livestock, pets, fish, and honeybees.

Some pesticides are more toxic than others. Be especially careful with toxic materials and take the precautions outlined here and on pesticide labels.

LEGAL CLASSIFICATIONS

Every pesticide is classified into one of two categories:

1. **General Use**—pesticides that can be purchased and applied by anyone
2. **Restricted Use**—pesticides that can be purchased only with a permit from a County Agricultural Commissioner

To obtain a permit to purchase restricted-use pesticides, you must be licensed by the California Department of Food and Agriculture as a pest control operator or otherwise certified as qualified to use such materials.

PRECAUTIONS

Follow recommendations carefully and exactly. Before handling any pesticide, read and understand all the information on the pesticide label. If you need help, ask for it. Know what to do in case of an accident. The authors of each chapter of this publication have given their recommendations for safely and effectively controlling specific pests. Follow those instructions.

Use the correct chemical. It is essential for your personal safety and for the safety of others, as well as of beneficial plants and animals, that when a specific chemical is recommended, you use no other. These chemicals have been tested and proved effective in controlling specific pests.

Use the correct amount. Use only the amount of chemical indicated on the container label. Adding "a little extra" will not control the best any better, and that extra amount may injure or burn the treated area or cause serious drift or other contamination problems.

Use at the proper time and temperature. When the directions say to apply the chemical at a specific time of day, or within a given temperature range, or at a certain stage of plant growth, they mean just that. Again, experiments and experience have shown that these specific application conditions give the safest, best control of the pest.

Use the recommended form of the chemical. Use only the formulation of the chemical recommended (wettable powder, dust, emulsifiable concentrate, granular, or gas fumigant) for good pest control. Plant injury (phytotoxicity) may result from the use of a different formulation of the same chemical.

Use the proper method of application. Applying the chemical in the prescribed manner is essential for effective control and safety. For example, if you are applying a chemical with a tractor-powered unit and the directions say to travel at a given speed, do just that. Increasing your speed will very likely mean that not enough chemical will be applied to a given area, resulting in poor or no pest control. If you travel slower than prescribed, you may deposit too much chemical and "burn" the turfgrass, leaving enough chemical residue to stunt plant growth. You could even prevent *any* further plant growth in the treated soil for a long time.

Do not combine chemicals unless recommendations say it is safe. Mixing certain chemicals can cause plant injury, reduced pest control, or clogged application equipment. Mixtures sometimes even increase the toxicity of the chemicals. Check the pesticide label or a compatibility chart before you combine pesticides.

HOW TOXIC ARE THE PESTICIDES?

You don't have to be a chemist or a toxicologist to understand approximately how toxic a pesticide is. Standard "signal words" appear in extra-large print on pesticide labels to indicate relative degrees of toxicity. Toxicity is classified as follows:

TOXICITY CLASS	SIGNAL WORD OR SYMBOL	APPROXIMATE TOXICITY
I	"Danger," "Poison," skull and crossbones	highly toxic
II	"Warning"	moderately toxic
III, IV	"Caution"	slightly toxic

> NOTE: Even though you may find no special signal words printed on the label of the pesticide you plan to use, always treat the chemical as a poison. Some pesticides with low toxicity ratings can accumulate to cause toxic reactions if used often and in large quantities.

SPLASHES AND SPILLS

As an applicator, you are unlikely to eat or drink a pesticide, except by accident. You are more likely, however, to splash or spill the chemical accidentally onto your skin or into your eyes. This is by far the most common means of poisoning for those handling pesticides. Therefore, your greatest concern should be to know how toxic the chemical will be to you if it gets on your skin or in your eyes—its "dermal toxicity." A pesticide's effects on you are determined mainly by

- how much gets on you
- how long it stays on you
- concentration
- toxicity
- formulation

A few drops of a highly toxic, highly concentrated liquid formulation can be very toxic to you if splashed or spilled onto your skin or into your eyes. If a pesticide splashes in your eyes, stop *immediately* and rinse your eyes with clean water for 10 to 15 minutes. If the chemical gets on your skin, *wash thoroughly* with soap and water. Then call your doctor to say what happened, giving the name of the chemical as printed on the package. *Remember, the longer you leave the chemical on your skin the more it will hurt you.* Liquid formulations containing solvents such as xylene, oils, and other petroleum hydrocarbons are absorbed by the eyes and skin much more rapidly than are dusts, wettable powders, or granular formulations.

Use special caution when pouring and weighing concentrated materials. They are most hazardous *before* dilution. However, even diluted chemicals can accumulate on your skin, hair, clothing, and shoes, so wash and change your clothing after application. If you are given a choice of chemicals to control a pest, select the one that is least hazardous to you and that will still control the pest.

Before handling any pesticide, put on clean protective clothing, a waterproof hat, boots, and a clear plastic face shield or goggles. Avoid breathing chemicals, especially dust and gas formulations. Wear a clean, well-fitted approved respirator when handling sprays and dusts. *Wear an approved gas mask when handling fumigants.* Check the label on the pesticide container for these and other precautions *before* you handle *any* pesticides. A list of approved pesticide respirators and gas masks should be available from your safety equipment dealer.

> NOTE: California regulations require that a closed pesticide mixing and loading system be used with all liquid Toxicity Class I pesticides.

WHAT TO DO AFTER APPLICATION

1. **Clean your application equipment** and put it back in good working order.
2. **Store your pesticides properly.** Use the original, labeled container—never store pesticides in or near food, feed, or drinking containers. Keep them in a locked, well-marked storage cupboard or building, never where they will get hot or freeze.
3. **Dispose of used pesticide containers properly.** Return 30- and 50-gallon drums to the original supplier, or have a drum reconditioner pick them up as soon as possible. Until you can get rid of them safely, store them in a place where no unauthorized person can get to them. Wash all non-returnable glass, metal, and plastic containers with soap and water, and rinse well. Make sure you pour the rinse water in a place where it will not contaminate water supplies or damage crops. Then break or puncture the containers so they cannot be reused. If you are a commercial grower, nursery operator, or applicator, you may be able to burn the containers.

Your local Air Pollution Control District can tell you whether burning is allowed in your area, and what restrictions apply. Stand out of the smoke fumes. Never burn containers that held hormone-type weed killer (e.g., 2,4-D). The fumes from these can drift, injuring plants many miles away.

KEEP GOOD RECORDS OF ALL APPLICATIONS

Immediately after the application, before you forget the details, record the pest treatment. Record the crop treated, the pest treated, the location and area treated, the time of day, the date of application, the exact name of the pesticide (including formulation), the amount used per gallon or unit of area treated, and the weather conditions.

> For more detailed information, refer to *Pesticide Application and Safety Training,* Publication 4070. For ordering information, write ANR Publications, 6701 San Pablo Avenue, Oakland, CA 94608-1239, or telephone (415) 642-2431.

2

Measurements, Calculations, and Sprayer Preparation

JOHN VAN DAM, TURFGRASS FARM ADVISOR, SAN BERNARDINO COUNTY
DAVID W. CUDNEY, EXTENSION WEED CONTROL SPECIALIST,
UNIVERSITY OF CALIFORNIA, RIVERSIDE

The areas of turfgrass that require treatment are generally much smaller than those treated in general agriculture. Consequently, in order to avoid overuse of the material, your measurements must be as precise as possible when you apply turfgrass pesticides. You must know how to calculate area measurements and how to determine pesticide applications for different-sized plots. The rates given in these illustrations are examples, not recommendations.

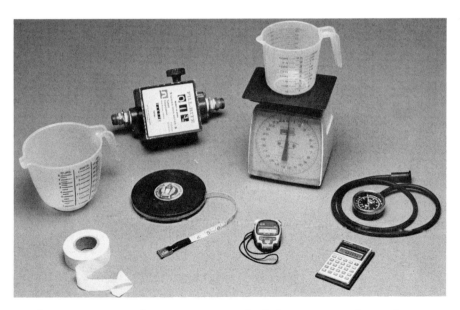

Simple equipment required to measure areas and calculate quantities of pesticides.

7

Two determinations must be made before treating any given area: (1) What is the size of the area to be treated? and (2) Precisely how much chemical will be used? Frequently, unsatisfactory control is blamed on the pesticide when, in fact, the fault lies with faulty calculations, either of the area to be treated or of the amount of pesticide to be applied, or both.

To ensure that measurements and calculations are as accurate as possible, you will need some simple equipment:

1. an easily read, well-marked 100-foot metal measuring tape
2. an inexpensive pocket calculator
3. a small scale that will weigh in ounces
4. one or more glasses, or a transparent plastic measuring device with appropriate divisions
5. a container of rinsing water
6. stakes

EXAMPLES AND FORMULAE

Determining the size of a given area can be simplified by dividing it into regular geometric shapes and using the formulae given in this section. Generally, any area (**A**) can be considered as a square or a rectangle. Odd extremities of the area can be visualized as measureable triangles or circles. For example, the fairways of a golf course can be visualized as rectangles; its tees as squares; and its greens, lakes, and water reservoirs as circles.

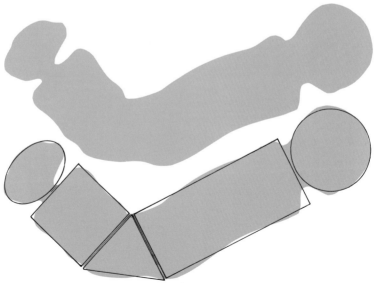

An irregular area reduced to simple geometric shapes.

SQUARE

Formula:
$A = s \times s$, where
s = side

Example: $s = 20$ ft
$A = 400$ sq ft (20×20)

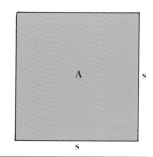

RECTANGLE

Formula:
$A = b \times h$, where
b = base (or length), and
h = height (or width)

Example: $b = 40$ ft
$h = 15$ ft
$A = 600$ sq ft (40×15)

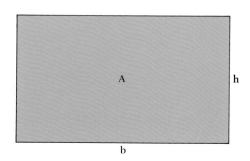

TRIANGLE

Where the geometric shape resembles a triangle, the formula is:

$$A = \frac{h \times b}{2} \text{ or } A = h \times b \times 0.5$$

In this formula, h is the height of the triangle and b is its base length. The area A is calculated by multiplying the height of the triangle times its base and dividing by 2, or multiplying by 0.5. The following illustrates the calculations for the area of a triangle:

Formulae: or
$A = \frac{h \times b}{2}$ $A = h \times b \times 0.5$, where
            h = height, and
            b = base length

Example: $h = 20$ ft
$b = 30$ ft
$A = 300$ sq ft $(20 \times 30 \times 0.5)$

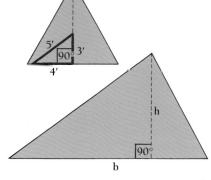

9

The height measurement *h* of the triangle must be perpendicular (at right angles) to the base *b*. A right angle (90°) is formed when any measurements having a ratio of 3, 4, and 5 are used to form a triangle. The right angle will be opposite the longest (5) side.

CIRCLE

When the geometric shape is a circle, the formula to determine its area is πr^2. The value of π is 3.14, and *r* is the radius of the circle. Another version of the same formula is being used increasingly: $A = 0.785 \times d \times d$. The letter *d* represents the diameter of the circle.

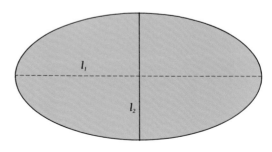

Formulae:
I. $A = \pi r^2$, where   or
    $\pi = 3.14$ II. $A = 0.785 \times d \times d$, where
    r = radius d = diameter

Examples:
I. $r = 8$ ft   II. $d = 16$ ft
   $A = 200.96$ sq ft $A = 200.96$ sq ft
   $(3.14 \times 8 \times 8)$ $(0.785 \times 16 \times 16)$

ELLIPSE

If the geometric shape resembles an ellipse rather than a circle, the formula $A = 0.785 \times l_1 \times l_2$ is used, with l_1 representing the length of the ellipse and l_2 the shorter length, or width.

Formula:
$A = 0.785 \times l_1 \times l_2$, where
l_1 = length of the ellipse, and
l_2 = width

Example: $l_1 = 15$ ft
        $l_2 = 5$ ft
        $A = 58.9$ sq ft $(0.785 \times 15 \times 5)$

IRREGULARLY SHAPED AREA

Method I. A very irregularly shaped area can be measured by establishing the longest line possible lengthwise through the center of the area. This length is considered the base (b). Numerous regularly spaced lines are then established perpendicular to this center line. The total number of perpendicular lines will depend upon how irregular the area's shape may be—the more irregular, the more lines it will require. The lengths of all these perpendicular lines are averaged to determine the overall height, and the area is then calculated as that of a rectangle.

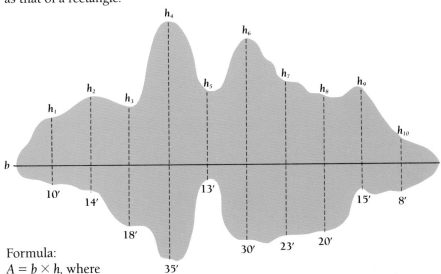

Formula:
$A = b \times h$, where
b = base (length of the area),
and
h = average of all heights h_1 to h_{10}

Example:
$h_1 = 10$ ft
$h_2 = 14$ ft
$h_3 = 18$ ft
$h_4 = 35$ ft
$h_5 = 13$ ft
$h_6 = 30$ ft
$h_7 = 23$ ft
$h_8 = 20$ ft
$h_9 = 15$ ft
$h_{10} = 8$ ft

Total 186 ft

$b = 128$ ft
$h = 18.6$ ft (186 ÷ 10)
$A = 2,380.8$ sq ft (128 × 18.6)

Method II. Another method for determining the size of an irregularly-shaped area, a golf green, for example, is to establish a point as near to the center of the area as can be estimated. From this point, as with a compass, measure distances at 10-degree increments to the edge of the irregularly shaped green. Then average the 36 measurements. That average becomes the radius of the circle. The diameter of the circle is found by multiplying its radius by 2. Then, compute the formula for a circle.

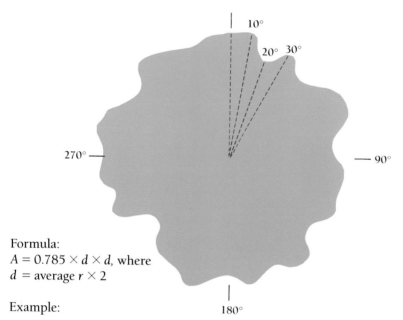

Formula:
$A = 0.785 \times d \times d$, where
d = average $r \times 2$

Example:

Degrees	Distance (ft): center to periphery
10 (r_1)	54.8
20 (r_2)	43.9
30 (r_3)	48.4
40 (r_4)	46.9
330 (r_{33})	41.5
340 (r_{34})	48.6
350 (r_{35})	51.0
360 (r_{36})	50.0
Total	1,980.0

$r = 55$ (1,980 ÷ 36)
$d = 110$ ($r \times 2$)
$A = 9,498.5$ sq ft (0.785 × 110 × 110)

PESTICIDE CALCULATIONS

Turfgrass recommendations generally are quoted on a per-1,000-square-feet basis. Label recommendations for pesticides usually are given on a per-acre basis, and the amounts of the pesticides to be applied are given in pints or pounds. To reduce label recommendations to the convenient equivalent amounts that are applicable to turfgrass applications you will have to make some simple calculations. Unless it is absolutely necessary to use precise amounts, you can convert the acreage figure by rounding off the square feet per acre (43,560) to 44,000 and dividing by 44 to obtain the equivalent per 1,000 square feet. That equivalent can then be reduced to its 100-square-feet equivalent by dividing by 10.

For example, if a recommendation called for 4.4 pounds of a dry weight material to be applied per acre, dividing that amount by 44 would give a figure of 0.1 pound per 1,000 square feet. With 16 ounces in a pound, this is equivalent to 1.6 ounces. To treat an area of 500 square feet, one-half of that amount would be needed. With 28.4 grams to an ounce, this is equivalent to 22.7 grams (28.4 g/oz × 0.8 oz). Several conversion tables are given below.

Recommendations also appear in terms of the active ingredient (ai) of a formulated pesticide. (Other ingredients are "inert," or nonactive.) Formulations vary. For dry formulations, the active ingredient is generally expressed as a percentage; for liquid formulations, it is expressed as the number of pounds of active ingredient per gallon of formulation.

To convert a recommendation given as pounds ai per acre (lb ai/A) for liquid formulations, calculate as follows:

Recommended rate (lb ai/A) ÷ lb/gal (see label) = gal/A.
Convert to the amount per 1,000 sq ft.

Example: If a recommendation calls for 2 lb ai/A of a 4 lb/gal formulation,
- Calculate 2 lb ai/A ÷ 4 lb/gal = 0.5 gal formulation/A.
- Then calculate 0.5 gal/A ÷ 44 = 0.011 gal/1,000 sq ft.
- Now, 0.011 gal/1,000 sq ft × 128 oz/gal = 1.4 oz of formulation required per 1,000 sq ft.

To convert a recommendation given as pounds ai per acre for dry formulations, calculate as follows:

Recommended rate (lb ai/A) ÷ (percentage ai in formulation ÷ 100) = lb/A. Convert to the amount per 1,000 sq ft.

Example: If a recommendation calls for 3 lb ai/A of a 75 percent wettable powder,
- Calculate 3 lb ai/A ÷ 75/100 = 3/0.75 = 4 lb formulation/A.

- Then calculate 4 lb formulation/A ÷ 44 = 0.09 lb/1,000 sq ft.
- Now, 0.09 × 16 oz/lb = 1.44 oz of formulation required per 1,000 sq ft.

NOW PREPARE YOUR SPRAYER

All the care you've taken to determine the area to be treated and the proper amount of material to apply is wasted unless you make sure to

1. check the sprayer:
    - Is the pump in working order (including seals, rollers, and the like)?
    - Are all the lines clear and in good shape?
    - Is the agitation system sufficient for the job you have in mind? (Wettable powders require the most vigorous agitation.)
    - Is the pressure regulator working properly to provide constant pressure?
    - Are all scale, dirt, and debris out of the system? Is the system clear?
    - Are proper screens in place at nozzles and pump?

2. check the nozzles and sprayer output:
    - Are the nozzles properly spaced on the boom?
    - Are the nozzles of uniform size?
    - Is the boom adjusted to proper height?
    - Did you run the sprayer with clear water?
    - Are the nozzle patterns and output uniform? (Check each nozzle's output for 1 minute; discard and replace nozzles that have poor patterns or that vary widely in their output.)

3. fill the tank with water, spray a known area at constant speed, and measure the amount used:
    Example: A 140- × 25-ft area is sprayed using 3.2 gal.
    - 140 ft × 25 ft = 3,500 sq ft.
    - 3.2 gal ÷ 3,500 sq ft = 0.914 gal/1,000 sq ft.
    - Using table 2, the spray volume would be slightly more than 40 gal/A.

4. make changes in the spray output in the following sequence:
    a. *Large changes.* Change the nozzle size. Example: To change from 40 gal/A to 80 gal/A, double the nozzle size.
    b. *Intermediate changes.* Change the travel speed. Example: To change from 40 gal/A to 50 gal/A, drive the rig slower.
    c. *Slight changes.* Vary the pressure. Take care to avoid high pressures, particularly with herbicides that pose drift hazards. (Most herbicide spraying with flat-fan nozzles should be at or below 40 psi.)

5. add herbicide to the spray tank:
   - Find the sprayer's output per 1,000 sq ft. In the example above, the sprayer applied 0.914 gal per 1,000 sq ft.
   - Add enough herbicide to treat 1,000 sq ft for each like amount of water you add to the spray tank. In this example, enough herbicide for 1,000 sq ft would be added with each 0.914 gal of water added to the tank.

   Example: If 2 oz of the formulation are to be applied per 1,000 sq ft, and the tank holds 30 gal, a full 30-gal tank will spray 32,820 sq ft (30 gal ÷ 0.914 gal/1,000 sq ft = 32.82), and 65.64 oz of formulation should be added to the 30-gal tank when it is filled (2 oz × 32.82).

6. check the sprayer periodically:
   - Make sure the output remains constant.
   - Do not vary the speed, spray height, or pressure once the sprayer is properly calibrated.

CONVERSION TABLES

Table 1.   English and metric (SI) units

Linear measure

1 inch (in)	= 1/12 ft		= 2.54 cm
1 foot (ft)	= 12 in	= 1/3 yd	= 0.305 m
1 yard (yd)	= 3 ft		= 0.914 m
1 land mile (mi)	= 5,280 ft	= 1,760 yd	= 1.609 km
1 centimeter (cm)	= 0.01 meter	= 0.394 in	
1 meter (m)	= 100 cm	= 3.281 ft	= 1.094 yd
1 kilometer (km)	= 1,000 m	= 0.621 mi	

Weight

1 ounce (oz)	= 1/16 lb	= 28.350 g
1 pound (lb)	= 16 oz	= 453.592 g
1 milligram (mg)	= 0.001 g	
1 gram (g)	= 1,000 mg	= 0.035 oz
1 kilogram (kg)	= 1,000 g	= 2.205 lb

Volume to weight (liquid water at 20°C at sea level)

1 pint (pt)	= 1.041 lb	= 0.473 kg
1 quart (qt)	= 2.086 lb	= 0.946 kg
1 gallon (gal)	= 8.330 lb	= 3.785 kg
1 milliliter (ml)	= 1 g	= 0.035 oz
1 liter (L)	= 1 kg	= 2.205 lb

Table 2. Equivalent rates of discharge

RECOMMENDED SPRAY RATE PER ACRE	EQUIVALENT RATE PER 1,000 SQ FT		
gal	gal	fl oz	L
10	0.227	29.1	0.861
20	0.455	58.2	1.722
30	0.682	87.3	2.584
40	0.909	116.4	3.445
50	1.136	145.5	4.307
60	1.364	174.5	5.165
70	1.591	203.6	6.027
80	1.818	232.7	6.888
90	2.045	261.8	7.749
100	2.273	290.9	8.611

Table 3. Equivalents for liquid measure

Unit	EQUIVALENT*								
	gal	qt	pt	c	fl oz	tbs	tsp	ml	L
gallon (gal)	1	4	8	16	128	256	768	–	3.785
quart (qt)	1/4	1	2	4	32	64	192	–	0.946
pint (pt)	1/8	1/2	1	2	16	32	96	–	0.473
cup (c)	1/16	1/4	1/2	1	8	16	48	–	0.237
fluid ounce (fl oz)	1/128	1/32	1/16	1/8	1	2	6	29.573	–
tablespoon (tbs)	1/256	1/64	1/32	1/16	1/2	1	3	14.706	–
teaspoon (tsp)	1/768	1/192	1/96	1/48	1/6	1/3	1	4.929	–
milliliter (ml)	–	–	–	–	0.034	0.068	0.203	1	0.001
liter (L)	0.264	1.057	2.114	4.228	–	–	–	1,000	1

*Partial English units appear as split fractions, except where they convert to metric (SI) units. Partial metric (SI) units appear as decimal fractions. Larger English units are converted only to liters, and smaller units only to milliliters.

3

Weed Control in Large Turf Areas

CLYDE L. ELMORE, EXTENSION WEED CONTROL SPECIALIST,
UNIVERSITY OF CALIFORNIA, DAVIS
W. B. MCHENRY, EXTENSION WEED CONTROL SPECIALIST,
UNIVERSITY OF CALIFORNIA, DAVIS
DAVID W. CUDNEY, EXTENSION WEED CONTROL SPECIALIST,
UNIVERSITY OF CALIFORNIA, RIVERSIDE

Different turfs serve many purposes, so they have different characteristics. Uniform texture is usually the most highly prized feature for golf courses and bowling green turf because of its good influence on the course of the ball. A deep green color is also desirable. These qualities require the most intensive management. However, where economy is an issue (in large areas such as parks), even crabgrass, if it makes up most of the turf, provides a pleasing enough overall appearance during the summer.

Irregular areas or patches of weeds that result from poor management make a turf unattractive and reduce its utility. A thick, vigorous, and competitive turf will resist any invasion of weeds.

Overwatering or underwatering, mowing too low or too high, low fertility, excessive wear, disease or insect damage, soil compaction, excessive shading—each one allows weeds to invade turf. Any condition that exposes the soil surface to additional light will tip the balance in favor of weeds.

Frequent, light irrigations encourage the germination of seeds of shallow-rooted weeds such as crabgrass and annual bluegrass. Such irrigation discourages deep-rooting turf and weakens its resistance to invasion. Mowing Kentucky bluegrass too short (shorter than 1½ inches) weakens the turf and encourages weed growth. Conversely, mowing bermudagrass too long (1 to 2 inches) decreases its competitiveness. Soil that is wet for long periods of time, often as a result of poor drainage, favors red sorrel, curly dock, nutsedge, and annual bluegrass.

CONTROL MEASURES

Good management practices can go far in controlling the weed problem. For instance, weeds recover more quickly from mowing than does turfgrass. Unless mowed frequently, weedy turf areas will develop a patchy appearance in just a few days. Annual bluegrass and ryegrass in the winter and tall fescue, dallisgrass, and buckhorn plantain in the summer are all weeds that require frequent mowing to prevent patchiness.

Efficient weed control can also be imposed with chemicals. Preplant fumigation of a turf site can kill bermudagrass, nutsedge, and other perennials, as well as all existing annual plants and many seeds. If weeds are already present, herbicides can still be introduced as a management tool to provide a decisive and rapid means of swinging the weed-grass competition in favor of the turf species. It should be noted, however, that diligent management must be maintained; otherwise, weeds will again encroach.

Consult leaflet 2589, *Selecting the Best Turfgrass*, for species adapted for your area; leaflet 2944, *Efficient Lawn Irrigation Can Help You Save Water*; leaflet 2587, *Mowing Your Lawn*; and leaflet 2586, *Lawn Aeration and Thatch Control*—all are published by the Division of Agriculture and Natural Resources, University of California, and are available from your local farm advisor, or from ANR Publications.

ORGANIZE YOUR CAMPAIGN

Know the enemy!
Are your weeds—

1) monocots (narrowleaf) or dicots (broadleaf)?
2) annuals (summer or winter), biennials, or perennials?

Improve your management practices!

1) improve turf species
2) use the correct mowing height
3) improve drainage
4) aerate and reduce compaction
5) improve irrigation practices to optimize turf growth
6) discourage practices that allow weeds to encroach

Choose your herbicide well!
Should it be—

1) selective or nonselective?
2) foliage applied or soil applied (or a soil fumigant)?
3) translocation or contact?

KNOW YOUR WEEDS

If a plant, whether a grass or a broadleaf, is growing where its presence is objectionable, it is a weed. Dichondra, a highly prized turf, may be considered an unwelcome intruder—a weed—in grass turf. Tall fescue is well suited as turf for areas receiving high wear, such as playing fields. But when it appears in a clump in Kentucky bluegrass or bermudagrass turf, destroying uniform texture, it becomes a weed.

The great majority of the 500 weed species in California are flowering plants, the higher form of plant life that produces seed. Some of the weeds are perennials, producing vegetative reproductive structures, too. For a well-composed weed control program, take into account the weed types that will be encountered and all control measures available; then select the most appropriate measures.

WEED CHARACTERISTICS

Monocots or dicots. Botanists recognize two natural divisions among plants, the monocotyledonae (monocots) and dicotyledonae (dicots). Certain weed killers are generally more effective against one or the other of these two plant groups. Therefore, it is important that anyone conducting weed control programs learn how to tell the difference. Your local county office of the University of California Cooperative Extension or Agricultural Commissioner's office can help you identify weed species. The local nurseries, too, should be acquainted with weeds common to your area.

Young seedlings emerge from the soil surface bearing small green seedling leaves (cotyledons) (fig. 1). The monocots have one seedling leaf, the dicots two. Once true leaves form, veins can be seen in the leaves. In the

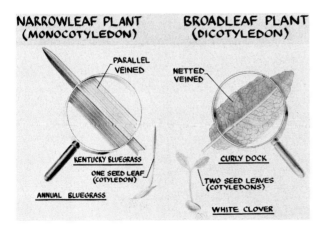

Fig. 1. Characteristics distinguishing monocots from dicots.

monocots, these veins almost always are parallel, extending from the leaf base to the tip. In dicots, the veins are "netted," and branch in several directions. Grasses, including Kentucky bluegrass, bamboo, iris, nutsedge, and palms are typical monocots. Maples, elms, oaks, chickweed, roses, and spotted spurge belong to the dicot group.

Many monocots, particularly the grasses, have narrow leaves, while most dicots have leaves that are more broad in relation to their length. Weed control professionals often call them narrowleaf and broadleaf weeds, but you must bear in mind that these terms are not literal and at times they may be inaccurate. For instance, buckhorn plantain, a dicot (broadleaf) and a common turf weed, has narrow leaves. Close inspection of the leaves shows that the secondary veins are netted, not parallel.

Annuals, biennials, and perennials. Plants are also classed as annuals, biennials, or perennials depending upon the number of years they live. Annuals grow from seeds, produce flowers, produce new seeds, and die in 1 year or less. There are summer annuals and winter annuals. Summer annuals germinate in the spring, grow to maturity during the summer, and die by fall or winter. Barnyardgrass, crabgrass, purslane, knotweed, and goosegrass are summer annuals. In southern California's mild climate, crabgrass sometimes lives into a second year. Winter annuals germinate in the fall or early winter and overwinter in a vegetative state (without flowering). In the spring they flower, mature a crop of seeds, and die. Examples of winter annuals are California burclover, common chickweed, common groundsel, and annual bluegrass.

Biennials have a life span of 2 years. The first year's growth is vegetative, usually as a rosette of leaves. During the second year a flower stalk arises and produces seeds, and the plant dies. Bristly oxtongue and little mallow (cheeseweed) are examples of biennial weeds.

Perennials live and produce seeds year after year. Many perennial plants, such as bermudagrass, also reproduce by vegetative means—when stolons or rhizomes break off of the mother plant, each new unit develops roots and leaves. Raking and pulling will not remove bermudagrass. You must remove or kill all of the underground rhizomes to eliminate this plant.

Perennial species in the seedling stage are as readily destroyed as annuals or biennials in the same stage. Less herbicide is required to kill perennials when they are attacked in the seedling stage. Once perennials become established, however, they can be difficult to destroy no matter what management practice you use.

Herbaceous perennials like nutsedge or bermudagrass have growing points below the soil surface. A contact herbicide such as cacodylic acid will completely destroy the aboveground shoots but will leave the underground, protected buds undamaged to grow again.

Annuals are the easiest and perennials the hardest plants to destroy. All plants are most vulnerable at the seedling stage, right after germination and before they become established. A crabgrass seedling in a thick, vigorous growth of turfgrass will die because of insufficient sunlight to support its adequate photosynthesis. In open, thin turf, crabgrass seedlings can be killed with appropriate herbicides.

KNOW YOUR HERBICIDES

The peculiar molecular structure of each herbicide makes it toxic to plants. Because plants are living organisms, their life processes are complex. Herbicides must find a way into plant tissues—via the roots or the shoot (the aboveground part of a plant)—in order to disrupt the plant's normal life processes.

When using any herbicide for the first time, apply it at the recommended rate on a limited area to make sure it is successful under your local conditions. If not used correctly, even selective herbicides can injure or kill desirable turf or ornamental plants. For this reason, herbicide users must follow recommendations for application rate, application timing, and any restrictions connected with drift.

Remember, plants are most susceptible to herbicidal action while in the seedling stage, but this susceptibility can apply to desirable plants as well as to weeds. Seedling turfgrass can be seriously stunted or killed with 2,4-D and other herbicides. Most soil-applied herbicides are used preemergence—before germination. With few exceptions, herbicides should be used only on established turf. As a rule of thumb, a new turf can be considered sufficiently established after it has grown enough to require its second mowing.

A single application of an appropriate preemergence, selective, soil-applied herbicide such as bensulide (Betasan) or DCPA (Dacthal) will control crabgrass in turf for an entire growing season. Once the crabgrass has emerged, a preemergent herbicide may not control it adequately or an excessive amount (many times greater than normal) of soil-applied herbicide may be required—so excessive that it may severely injure or kill the turf.

To control emerged crabgrass and preserve necessary selectivity, use a compound like DSMA, which acts on the foliage but has little or no soil activity. Repeated applications of DSMA, applied 7 to 10 days apart, are often required to rid the turf of crabgrass.

Another approach is to purchase a mixture of two crabgrass herbicides—a preemergence, soil-applied herbicide and a postemergence, foliar compound for use after crabgrass is established. The foliage-active ingredient kills the crabgrass that is present, and the selective, soil-applied herbicide prevents later-germinating crabgrass from becoming established.

Foliage-active herbicides like 2,4-D, triclopyr, or MSMA penetrate the leaves and stems and move within the vessels of the plant's vascular system, down to the roots and underground buds. This movement within the plant is called translocation. Often, a respray will be necessary when new shoots form on perennials, in order to achieve translocation in a sufficient concentration to destroy all the buds. A wetting agent (surfactant) is often added to foliar sprays to help penetrate leaves. Certain herbicides, such as MSMA, are mainly effective in destroying monocots; others, like 2,4-D, are more effective against dicots. This preferential toxicity provides much of the selectivity that has made herbicides so useful. There may be several herbicides for a specific weed problem, each differing in selectivity, weed toxicity, longevity of control, convenience of use, and price.

Formulations. Herbicide recommendations suggested by the University of California are given in units of active ingredient (a.i.). This is necessary because herbicides are sold in many different concentrations. Herbicides, for the most part, are marketed either as dry materials or as liquids (table 1).

With dry formulations, the strength of the active ingredient is given on the label as a percentage. Dacthal (DCPA) as a wettable powder is 75 percent active. If 15 pounds of active ingredient were called for, 20 pounds of commercial product would be required. Oxadiazon is a 2 percent dry granule formulation. If 2 pounds of active ingredient were called for, 100 pounds of the commercial product would be needed. Pendimethalin may be purchased as a granule and spread on the turf or as a water dispersable granule (WDG) that is mixed in water and sprayed on the turf. Always read the label for formulation, mixing, and treatment directions.

Table 1. Herbicide Formulations and Application Methods

FORMULATION	EXAMPLE	APPLICATION METHOD
DRY		
Soluble powder	DSMA	dissolve in water, and spray
Wettable powder	DCPA, diphenamid, napropamide	suspend in water by stirring, then spray (requires agitation to prevent settling out in a tank)
Water-dispersible granule	pendimethalin	suspend in water by stirring, then spray
Granule	benefin, DCPA, oxadiazon, benefin plus oxadiazon, benefin plus trifluralin, benefin plus oryzalin	apply dry directly from the package (should be applied with a drop spreader or a rotating-disc spreader)
LIQUID		
Soluble concentrate	fluazifop, metham, MSMA, sethoxydim	dissolve in water, and spray
Emulsifiable concentrate	bensulide, 2,4-D ester, bromoxynil, dicamba, mecoprop, triclopyr, several combinations	emulsify with water by vigorous stirring, then spray (requires frequent agitation to prevent separation of water from concentrate)

The active ingredient in a liquid formulation is expressed in pounds of active ingredient per gallon. A gallon of 2,4-D amine usually contains 4 pounds of active 2,4-D (expressed as the acid equivalent). To apply 1 pound of active 2,4-D per acre, ¼ gallon (1 quart) of the formulation would be needed.

Selective or nonselective. There are several ways to narrow down the choice of herbicide for a particular weed problem. The most useful first step is to determine which herbicides are selective and which are nonselective. The herbicide 2,4-D will control many broadleaf weeds with little or no effect on grass turf when used at recommended rates; it destroys weeds selectively. However, application rates of 2,4-D that can be used selectively in Kentucky bluegrass or bermudagrass turf may injure bentgrass turf; therefore, lower rates must be used to preserve selectivity. The grass killer fluazifop will control weedy grasses selectively in dichondra when used at recommended rates, but is nonselective if applied on grass turf. The herbicide formulation, application rate, stage of growth of desirable plantings to be preserved, temperature during and following application, and soil moisture status all affect selectivity in varying degrees. Thus, again, selectivity is relative and depends upon the use made of the herbicide. Table 2 shows that several herbicides can be used either selectively or nonselectively.

Foliage- or soil-applied. The next step is to determine whether the herbicide should be used as a foliage-spray or soil-applied treatment. Some of the spray applied as a foliage treatment will, of course, reach the soil; similarly, soil-active sprays will wet the leaves. However, foliage-applied herbicides act primarily upon the leaves or shoots of the plant. These herbicides must not be washed from the leaves with water after application. Soil-applied herbicides are taken up primarily by the roots and must be carried into the root zone by rain or irrigation in order to become effective. Soil fumigants such as metham are in reality short-term soil-applied compounds that require special application techniques such as use under a vaporproof covering like polyethylene.

Translocation- or contact-type. The final step (if the herbicide is foliar-applied) is based on the action of the chemical on the plant. Translocation-type herbicides penetrate the leaves and stems, move in the vascular system, and eventually reach and kill the growing points in both the aboveground and belowground parts. Herbicides that translocate include fluazifop, triclopyr, glyphosate, and MSMA. Some translocated herbicides such as dicamba and triclopyr also have some soil activity and can be taken up by roots. The less mobile foliar herbicides like cacodylic acid kill only the plant tissues touched by the spray. Movement within the plant beyond the point of contact is relatively limited. These are called contact herbicides.

Table 2. Application and Action of Selective and Nonselective Turf Herbicides*

HERBICIDE	SELECTIVE FOLIAGE-APPLIED CONTACT	SELECTIVE FOLIAGE-APPLIED TRANS-LOCATED	SELECTIVE SOIL-APPLIED	NONSELECTIVE FOLIAGE-APPLIED CONTACT	NONSELECTIVE FOLIAGE-APPLIED TRANS-LOCATED	NONSELECTIVE SOIL FUMIGANT
AMA (amine methanearsonate)		■				
Ammonium methylarsonate		■				
Benefin (Balan)			■			
Bensulide (Betasan, Presan)			■			
Bentazon (Basagran)	■					
Bromoxynil	■					
Cacodylic acid				■		
Chlorfluoranol			■			
CMA (calcium methanearsonate)		■				
Dazomet (Basamid)						■
DCPA (Dacthal)			■			
Dicamba (Banvel)		■			■	
Dichlorprop		■				
Diphenamid (Enide)			■			
DSMA		■				
Endothal (Endothal Turf Herbicide)	■			■		
Fluazifop (Fusilade 2000)		■				
Glyphosate					■	
Mecoprop (MCPP)		■				
Metham (Vapam, VPM)						■
MSMA		■				
Napropamide (Devrinol)			■			
Oryzalin			■			
Oxadiazon (Ronstar G)			■			
Pendimethalin			■			
Pronamide (Kerb)			■			
Sethoxydim (Poast)		■				
Siduron			■			
Triclopyr (Turflon)		■				
2,4-D		■			■	

*This listing classifies available herbicides based on the ways in which they are used. It does not necessarily constitute a recommendation for such use.

HERBICIDE APPLICATION

Care and accuracy in applying herbicides is very important in all weed control jobs. Excessive herbicide rates are wasteful and may destroy or seriously injure desirable plantings. Selectivity is relative; exceed the recommended application rate, and you may lose that selectivity. Insufficient application, on the other hand, usually results in failure or incomplete weed control. Don't depend on makeshift spray equipment to apply herbicides accurately.

When using controlled droplet applicators (CDA), be sure to maintain normal operating revolutions per minute. By increasing the revolutions per minute, you will increase the number of fine droplets and increase the chance for spray drift.

Spray rig pressures should be kept at a minimum, generally 30 pounds or less per square inch (psi) to prevent drift and to confine the herbicide to the immediate target area. The higher the pressure applied at the nozzles, the smaller the spray droplet size. Small droplets remain airborne and can move or drift over great distances and damage susceptible and expensive ornamentals. All spray or granular applications should be done under calm air conditions to reduce drift.

Various nozzle tips can be used on boom sprayers (fig. 2). With most herbicide applications, a flat-fan tip gives even distribution for soil-applied sprays and can also be used for foliage-applied sprays. The solid-cone and hollow-cone nozzle tips are used for spraying other pesticides, such as insecticides or fungicides.

Figure 3 illustrates flat-fan nozzles spaced at proper intervals (*A*) with the boom height adjusted to provide double coverage for uniform application rates (zone 2). Each parallel spray swath should be overlapped to cover zones 1 and 3 at the ends of the boom. The same procedures should be used when covering small areas with a single flat-fan nozzle (*B*).

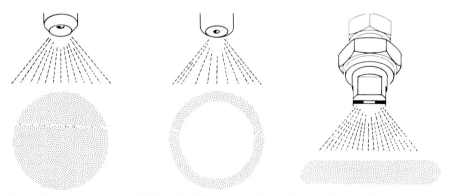

Fig. 2. Spray patterns available with different nozzles: solid cone (left), hollow cone (center), and flat fan (right). An even distribution of the material over the area being treated is necessary at all times.

Apply herbicides uniformly. Application rates are usually given as a certain weight (in dry formulations) or volume (in liquid formulations) per acre or per 1,000 square feet. Such recommendations assume that every square foot of ground or turf covered will receive a uniform amount. On extensive turf areas, such as parks and golf courses, this is accomplished most effectively with a powered sprayer equipped with evenly spaced nozzles on a boom (fig. 4). Nearly all weed-control spraying uses nozzles that produce a flat-fan spray pattern.

When applying spray by hand on a small area, walk at a uniform speed and stop to pump up the sprayer as soon as you notice a change in the spray pattern. If you allow the pressure to drop too far, the spray swath will narrow and less material will be applied. Avoid hesitation; otherwise, unwanted spray will be applied on clumps of weeds. Keep moving at a constant speed over every square foot, whether you see weeds or not. To maintain a uniform mixture, occasionally shake the tank vigorously if you are applying a wettable powder or an emulsifiable formulation.

Some companies market granular herbicides—often formulated with fertilizer—that can be applied conveniently and uniformly over a small area direct from the package with a lawn spreader (fig. 5). However, granular formulations are more expensive per pound of active herbicide. Whether applying the herbicide wet or dry, you can actually obtain more uniform application by applying one-half of the required herbicide over the entire area to be treated in one direction (say, north-to-south), and then completing the application by applying the remaining half of the herbicide over the entire area in the perpendicular direction (east-to-west).

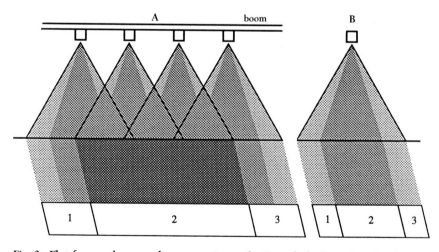

Fig. 3. Flat-fan nozzles spaced at proper intervals (A) with the boom height adjusted to provide double coverage for uniform application rate (zone 2). Each parallel spray swath should be overlapped to cover zones 1 and 3 at the ends of the boom. The same procedure should be used when covering small areas with a single flat-fan nozzle (B).

Nonselective sprays can move to desirable turf. When spraying glyphosate on weeds, a small amount may adhere to the rubber boots of application personnel or on the hoses used during application. Take care to wash the material off or to walk on soil before walking on or dragging hoses over desirable turf. Care should also be taken in applying irrigation water after the herbicide application. Do not overirrigate—the excess water runoff may move the herbicide to an untreated area, resulting in damage to desirable turf.

Do not use 2,4-D, mecoprop, triclopyr, dicamba, or combinations of these materials in spray equipment that is also used to apply insecticides or fungicides to ornamentals. The very small amounts of these materials that remain in the equipment, even after several washings, can injure or kill ornamentals.

Spot-spraying individual weeds with herbicides is probably the most useful method for very small plots. A foliage-applied herbicide such as 2,4-D is mixed with water at the recommended concentration and applied in a short burst of spray while holding the nozzle directly above the weed. Apply just enough spray to thoroughly wet the leaves—but no more. Excessive drenching can kill the turfgrass in the sprayed spot. Stoloniferous turfgrasses, particularly bentgrass and bermudagrass, are easily injured by spot treatments. Unless the weeds are widely scattered, it is safer and only slightly more expensive to spray a larger area with equipment that provides precise, uniform application.

Fig. 4. A powered sprayer is excellent for applying herbicides to large areas such as parks and golf courses.

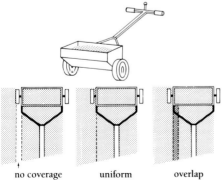

Fig. 5. When using this type of granule applicator, bands of chemical must meet to prevent gaps or overapplied areas.

Spot treatments are useful for nonselective, foliage-applied herbicides. For example, glyphosate can be used to control individual clumps of tall fescue, ryegrass, or bermudagrass in grass turf. Use a very low pressure to prevent excessive application. Localized applications can be useful around sprinkler heads, in sidewalk cracks, and along lawn edges. Use the same protective clothing as you would use for a large-scale application.

You can apply spot treatments of glyphosate for weedy grasses or of 2,4-D for broadleaf weeds using a paint brush or a small rag or sponge fixed to the end of a stick, daubing each individual weed. Use the same herbicide and wetting agent concentration in water as is recommended for spray applications. This application technique will kill dandelions in dichondra, using 2,4-D, even though dichondra is very sensitive to broadleaf weed herbicides such as 2,4-D. Some herbicides are packaged in convenient ready-to-use aerosol spray cans for spot applications in home turf.

Characteristics of Turfgrass Weeds, and Their Annual Periods of Growth and Control

NARROWLEAF WEEDS	GROWTH TYPE	METHOD OF PROPAGATION
Annual bluegrass—*Poa annua*	A,P	S
Barnyardgrass—*Echinochloa crusgalli*	A	S
Bentgrass—*Agrostis* spp.	P	S,St
Bermudagrass—*Cynodon dactylon*	P	S,R,St
Crabgrass—*Digitaria sanguinalis* or *D. ischaemum*	A	S
Dallisgrass—*Paspalum dilatatum*	P	S,R
German velvetgrass—*Holcus mollis*	P	S,R,St
Goosegrass—*Eleusine indica*	A	S
Italian ryegrass—*Lolium multiflorum*	A,P	S
Kikuyugrass—*Pennisetum clandestinum*	P	S,R,St
Meadow fescue—*Festuca elatior*	P	S
St. Augustinegrass—*Stenotaphrum secundatum*	P	St
Yellow foxtail—*Setaria glauca*	A	S
Yellow nutsedge—*Cyperus esculentus*	P	S

A = Annual S = Seed
B = Biennial St = Stolon
P = Perennial R = Rhizome
Rs = Rootstocks

- Period of active growth
- Selective preemergence control
- Selective postemergence control

Characteristics of Turfgrass Weeds, continued

BROADLEAF WEEDS	GROWTH TYPE	METHOD OF PROPAGATION	JANUARY	FEBRUARY	MARCH	APRIL	MAY	JUNE	JULY	AUGUST	SEPTEMBER	OCTOBER	NOVEMBER	DECEMBER
Birdseye pearlwort—*Sagina procumbens*	P	S												
Birdseye speedwell—*Veronica persica*	A	S												
Black medic—*Medicago lupulina*	A,P	S												
Broadleaf plantain—*Plantago major*	P	S												
Buckhorn plantain—*Plantago lanceolata*	P	S												
California burclover—*Medicago polymorpha*	A	S												
Chickweed—*Stellaria media*	A	S												
Common purslane—*Portulaca oleracea*	A	S												
Creeping spurge—*Chamaesyce serpens*	P	S												
Creeping woodsorrel—*Oxalis corniculata*	P	S,Rs												
Cudweed—*Gnaphalium chilense*	A,B	S												
Curly dock—*Rumex crispus*	P	S												
Cutleaf geranium—*Geranium dissectum*	A	S												
Dandelion—*Taraxacum officinale*	P,B	S												
Dichondra—*Dichondra micrantha*	P	S,St												
English daisy—*Bellis perennis*	P	S												
Field bindweed—*Convolvulus arvensis*	P	S,R												
Field madder—*Sherardia arvensis*	A	S												
Healall—*Prunella vulgaris*	P	S,Rs												
Henbit—*Lamium amplexicaule*	A	S												
Little mallow—*Malva parviflora*	B	S												
Mouseear chickweed—*Cerastium vulgatum*	A,P	S,Rs												
Pennywort—*Hydrocotyle umbellata*	P	S,Rs												
Prickly lettuce—*Lactuca scariola*	A	S												
Prostrate knotweed—*Polygonum aviculare*	A	S												
Red sorrel—*Rumex acetosella*	P	S												
Scarlet pimpernel—*Anagallis arvensis*	A	S												
Southern brassbuttons—*Cotula australis*	A	S												
Spotted catsear—*Hypochoeris radicata*	P	S												
Spotted spurge—*Euphorbia maculata*	A	S												
Spurweed—*Soliva sessilis*	A	S												
Wartcress—*Coronopus didymus*	A	S												
White clover—*Trifolium repens*	P	S												
Yarrow—*Achillea millefolium*	P	S,Rs												

ANNUAL BLUEGRASS

Poa annua L.

Annual; tufted, light green, forms mat when allowed to mature without cutting. Stems are flattened, and leaf blades are short and smooth. Seeds continue to form even under extremely close mowing. May root at the lower nodes. Usually found in cool, frequently watered area.

BARNYARDGRASS
[WATER GRASS]

Echinochloa crusgalli (L.) Beauv.

Annual; stout stems, branching at base, forms mat or clumps when mowed. Leaves are smooth with prominent midrib. Seed heads are panicles, and seeds are about the size of millet (birdseed). Normally found in poorly managed turf of low fertility. Commonly called "watergrass."

BENTGRASS

Agrostis spp.

Perennial; fine-bladed, spread by seed and stolons. Roots easily and forms stems at every node. Develops dense sod patches, normally much finer than other turf. Commonly used as a turfgrass.

NARROWLEAF WEEDS

BERMUDAGRASS
Cynodon dactylon (L.) Pers

Perennial; with rhizomes and stolons. Stolons (runners) are many-jointed and root at the nodes. Stems are smooth and wiry. Fine leaves are pointed at the end, and have a conspicuous tuft of hair at the base. Does not grow well in shade. Seed heads resemble those of crabgrass. Commonly used as a turfgrass in southern and central California. Details: Middle: shoot leaf base, with sparse hairs. Bottom: seed head, showing single point of origin.

LARGE CRABGRASS (HAIRY CRABGRASS)

Digitaria sanguinalis (L.) Scop.

Annual; spread by seed, and to a minor extent by rooting from lower swollen nodes of stems. Has low-bunching leafy grass with leaves that are larger and hairier than those of smooth crabgrass. Light green to yellowish green in color. Often heavy in overwatered turf.

SMOOTH CRABGRASS

Digitaria ischaemum (Schreb.) Muhl.

Annual; similar to hairy crabgrass, except that the leaves and leaf sheaths are smooth. Leaves are longer and narrower than those of hairy crabgrass. (Shown here in dichondra.)

DALLISGRASS

Paspalum dilatatum Poir.

Perennial; bunch-type growth, leafy at the base. Rhizomes are very closely jointed, appearing almost scaly. Seed heads are sparsely branched on long stems. Detail: seed head and circular, flat seeds.

NARROWLEAF WEEDS

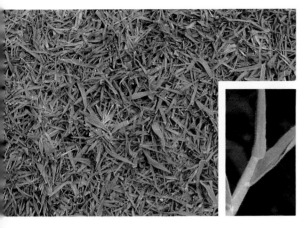

GERMAN VELVETGRASS

Holcus mollis L.

Perennial; vigorous, slender rhizomes. Leaves are velvety, somewhat grayish. Purplish lines at base of stems are common. Visible as light velvety clumps in turf. Detail: leaf base with many short, fine hairs on leaves and leaf sheath.

GOOSEGRASS

Eleusine indica (L.) Gaertn.

Annual; low rosette, mat-forming, stems compressed. Appears as silvery, pale green clump. Flower stalks are short, stout, and compressed. Seed heads are similar to those of dallisgrass, but short and stiff. Normally found in compacted areas or areas of heavy wear. Produces seed even under close mowing.

ITALIAN RYEGRASS

Lolium multiflorum Lam.

Annual or short-lived perennial; robust plant with long, narrow, glossy leaf. Spikes are long, with spikelets attached on alternate sides. Recognizable in lawns by its glossy appearance; forms clumps. Often used in inexpensive lawn seed mixtures.

KIKUYUGRASS

Pennisetum clandestinum Hochst.

Perennial; vigorous, thick rhizomes and stolons. Seeding is very sparse, inconspicuous. Mowing stimulates seed production. Leaves are pointed on the ends—similar to Bermudagrass, except coarser. Wiry, it is often mistaken for St. Augustinegrass. Found in south and central coast. Detail: leaf base with hairs on leaves and leaf sheath.

ST. AUGUSTINEGRASS

Stenotaphrum secundatum (Walt.) Kuntze

Creeping perennial; thick, flattened stolons, spreading leaves have boat-shaped tips. Leaves and stems are smooth and hairless. Flower shoots are 4 to 12 inches tall. Sometimes used as a turfgrass.

TALL FESCUE

Festuca arundinacea Schreb.

Perennial; forms clumps with upright leaves. Leaf blades are wide, coarse with rasplike margins, and radiate from a central clump. Flower stalks lie flat during mowing, resulting in ragged-looking turf. Used for turf in playgrounds and parks. Meadow fescue (Festuca elatior L.) has similar clumping characteristics; it is included in some seed mixtures.

NARROWLEAF WEEDS

WILD BARLEY (COMMON FOXTAIL)

Hordeum leporinum Link.

Annual; occurs as clumps, principally in newly seeded turf. The leaves are smooth, dull green. Seed heads are squirrel- or foxtail-shaped, and often remain after mowing.

YELLOW FOXTAIL

Setaria glauca (L.) Beauv.

Summer annual; erect, 1- to 2-foot stems. Leaves are flat, some with a spiral twist, and with long hairs on the upper surface near the base. Spikes are dense and erect; spikelets have five or more slender bristles. Found in turf, cultivated fields, and roadsides. Detail: seed head with bristles and seeds.

YELLOW NUTSEDGE

Cyperus esculentus L.

Perennial; spreading by seeds and tubers. Looks like grass, but with triangular stems. Nutsedge also has three tiers of leaves, whereas grasses have two tiers. Leaves are shiny, dark green, stiff, and upright. Grows faster than turfgrasses. Underground tubers form from the scaly rhizomes. Nutlets have almond taste.

BIRDSEYE PEARLWORT

Sagina procumbens L.

Perennial; stems prostrate, forms mat, rooting at nodes. Commonly found in coastal areas. Looks like moss.

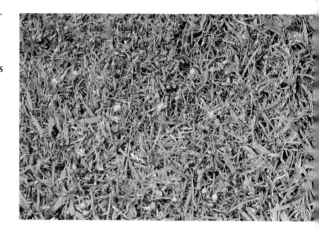

BIRDSEYE SPEEDWELL (BYZANTINE SPEEDWELL)

Veronica persica Poir.

Annual; stems 4 to 16 inches tall, leaves roundish or oval. Has small, deep blue flowers with white center on stalks ⅜ to 1 inch long. Plant is covered with hairs.

BLACK MEDIC

Medicago lupulina L.

Annual or sometimes perennial; four-angled stems are hairy and branch at the base. Branches are prostrate and spreading. Flowers are small and yellow, in long, dense heads. Detail: leaflets and flower head with small yellow flowers.

BROADLEAF WEEDS

BRISTLY MALLOW
Modiola caroliniana (L.) G. Don.

Low perennial; spreading stems 6 to 18 inches long. Leaves are round with coarse-cut margins 1 to 1½ inches across. May look similar to cutleaf geranium. Flowers are dull red, round, and less than ½ inch wide in summer. Found throughout most of California. Detail: *flower and leaf.*

BRISTLY OXTONGUE
Picris echioides L.

Coarse, rough biennial; 4 to 6 inches tall in turf. Leaves are 2 to 6 inches long, ½ to 1½ inches wide, rough and hairy on upper and lower surfaces. Flower heads are yellow, about ½ inch broad, and occur in clusters near the top of the stem. Flowers appear from June to December. Detail: *bristly leaf surface.*

BROADLEAF PLANTAIN
Plantago major L.

Perennial; leaves are large, 3 to 6 inches long, and smooth. Conspicuous flower stalks 3 to 6 inches long. Roots are fibrous and shallow. May be distinguished from buckhorn plantain by the broader leaf and shorter flower stalks.

BUCKHORN PLANTAIN
Plantago lanceolata L.

Perennial; long, narrow leaves with parallel veins, 3 to 12 inches long, slightly hairy, with brownish hairs surrounding the leaf base. Growth habit resembles that of broadleaf plantain. Seed stalks are much longer than leaves.

CALIFORNIA BURCLOVER
Medicago polymorpha L.

Annual; smooth prostrate stems. Leaflets have whitish and red spots scattered over the surface when young. Small flowers are yellow. Seed pod of "bur" is twisted spirally with spiny protrusions. Stipules at the base of leaflets are deeply divided.

CHICKWEED
Stellaria media (L.) Vill.

Annual; slender, many-branched stems with line of white hairs along one side. Leaves are smooth and pointed. Grows in cool weather or, in winter, in sheltered areas. Detail: flowers and pointed leaves.

COMMON PURSLANE
Portulaca oleracea L.

Annual; prostrate, fleshy stems, flowers are pale yellow. Leaves are fleshy and succulent.

CREEPING SPURGE
Chamaesyce serpens H. B. K. (Small)

Perennial; smooth, prostrate, spreading and rooting from the nodes. Leaves are almost round, reaching ¼ inch in length. Light green color, with no spots. Small maroon glands are often mistaken for seed pods. Often found at turf edge or in open ground in central valleys and southern California. Detail: stems, leaves, flowers, and seed pods.

CREEPING WOODSORRELL (YELLOW OXALIS)

Oxalis corniculata L.

Perennial; with running rootstocks. Prefers shade. Leaves resemble those of clover, green or reddish color with sour taste. Flowers are small and yellow. Detail: *purple leaf form with small yellow flower.*

CUDWEED (COTTON BATTING PLANT)

Gnaphalium chilense Spreng.

Annual or biennial; white, silk-hairy plant with long, narrow leaves with round tips at base and more pointed tips at the top of the plant. Small inconspicuous flowers appear in clusters.

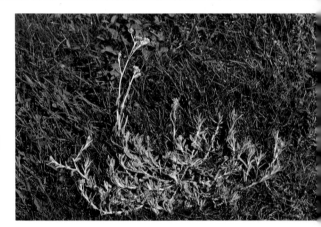

CURLY DOCK

Rumex crispus L.

Perennial; growing from a large, brownish taproot. Normally found as a rosette in turf, but will grow to 2 or 3 feet. Leaves are large, reddish green in color, curly, and wavy along the margins.

BROADLEAF WEEDS

CUTLEAF GERANIUM
Geranium dissectum L.

Annual; freely branched, remains prostrate in turf. Leaves are dissected into narrow leaf divisions. Small, purple flowers are conspicuous.

DANDELION
Taraxacum officinale Weber

Perennial; stemless leaves at base. Flower stalks arise from the base. Has conspicuous yellow flowers. At maturity, flower becomes a white fluffy ball of a seed head. The plant grows from a single brownish taproot.

DICHONDRA
Dichondra micrantha Urb.

Perennial; low, creeping habit, roots freely at nodes, and forms dense mat. Leaves are kidney-shaped and from ¼ to 1½ inches wide. Flowers are inconspicuous.

ENGLISH DAISY
Bellis perennis L.

Perennial; low-growing, with oval basal leaves. Has prominent upright flower stalks throughout season with showy white or pinkish daisylike flowers.

FIELD BINDWEED
Convolvulus arvensis L.

Perennial; deep-rooted, with prostrate or twining stems and arrow-shaped leaves. Flowers are white or pink and conspicuous. Plant can take many forms and shapes.

FIELD MADDER
Sherardia arvensis L.

Annual; slender, square, hairy stems, low-growing, forms mat. Leaves grow in whorls and have a strong odor. Flowers are small and blue or pinkish in clumps at ends of the stems. Detail: stems, leaves, and flowers.

BROADLEAF WEEDS

HEALALL (SELFHEAL)
Prunella vulgaris L.

Perennial; gray, hairy or smooth green plant, with spreading rootstocks. It is usually found in patches. There is also a silvery, hairy form. Detail: flower head.

HENBIT
Lamium amplexicaule L.

Annual; several upright four-angled stems, rooting at lower nodes. Rounded, coarsely toothed leaves are whorled up the stem. Has an irregular purple flower. The lower lips of the flower are spotted. A difficult weed to control.

LITTLE MALLOW (CHEESEWEED)
Malva parviflora L.

Annual or very often a biennial; stems somewhat low spreading. Leaves are roundish and broad, with a red spot at base of the blade. Normally found in poorly managed or new turf.

LOW AMARANTH
Amaranthus deflexus L.

Annual; low, prostrate, with the end of the stem turning upward. Stems are 12 to 18 inches long. Leaves are short-petioled, oval, ¼ to 1 inch long and about ½ inch broad. Flowers are inconspicuous, cluster densely at the stem ends, and appear from May to November. Another pigweed, prostrate pigweed, is very similar, except that it forms a small, dense mat.

MOUSEEAR CHICKWEED
Cerastium vulgatum L.

Perennial; prostrate, forms dense patches, sticky-hairy stems and leaves, leaves rounded on end. Flowers are small, white, and inconspicuous. Detail: hairy leaves and small, white flower.

PENNYWORT
Hydrocotyle umbellata L.

Perennial; creeping, rooting at nodes of slender rootstocks. A troublesome weed in southern California. Round leaves are approximately ½ inch in diameter with wavy margins that distinguish it from dichondra.

BROADLEAF WEEDS

PROSTRATE KNOTWEED
Polygonum aviculare L.

Annual; prostrate, forms circular mats, found in hard-worn turf areas of high traffic. The slender, wiry stems do not root at the nodes. Leaves are bluish green and smooth, without the purple spotting found on spotted spurge. Flowers are white and inconspicuous. Detail: *stems, leaves, and flowers.*

RED SORREL (SHEEP SORREL)
Rumex acetosella L.

Perennial; similar to curly dock, arrow-shaped leaves. Root is a slender running rootstock, reddish in color.

SCARLET PIMPERNEL
Anagallis arvensis L.

Annual; low-growing, branched, with four-angled stems. Leaves are opposite or in whorls of three. Flowers are salmon-colored, about ¼ to ⅓ inch in diameter, and open only under a clear sky.

BROADLEAF WEEDS

SOUTHERN BRASSBUTTONS (AUSTRALIAN BRASSBUTTONS)

Cotula australis (Sieber) Hook. f.

Annual; strong-scented, low-growing. Resembles Soliva or wartcress. The flowers are small and yellow, and the leaves are deeply toothed and covered with fine hairs.

SPOTTED CATSEAR (HAIRY CATSEAR)

Hypochoeris radicata L.

Perennial; thick, fleshy taproot, normally found as a rosette in turf. Leaves are lobed or sawtoothed, with coarse yellow hairs on upper and lower surfaces. Flowers are yellow, and each stem may have several.

SPOTTED SPURGE

Euphorbia maculata L.

Annual; prostrate, stems form circular mat from a single taproot, nonrooting at the nodes, aggressive. Milky sap in leaves, which may be green or reddish, but are easily identified by the red spot on the upper center of the leaflet.

BROADLEAF WEEDS

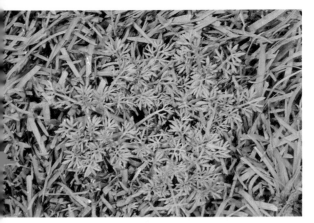

SPURWEED

Soliva sessilis R. & P.

Annual; small, with dissected leaves. Foliage is covered with fine hairs. Seed, when formed, appears in a small, three-spined bur that you can feel readily when you press a clump with your hand. Detail: inconspicuous flower head in the leaf axil with small spines.

WARTCRESS

Coronopus didymus (L.) Sm.

Annual; low-growing, stong-smelling, closely resembles Soliva. Seeds are borne in two-lobed, wrinkled pods. Detail: leaves, flowers, and seed pods.

WHITE CLOVER

Trifolium repens L.

Perennial; low-growing, rooting at the nodes. Has white, sometimes pinkish flowers, trifoliate leaf with distinctive white mark. Detail: leaflets with white, crescent-shaped mark.

WHITESTEM FILAREE

Erodium moschatum (L.) L'Her.

Annual; leaves form a low rosette. Stems are generally whitish. Leaves are up to 10 inches long, with several toothed leaflets. Flowers are small, bluish purple. Seed pods form a stork's bill. Detail: leaf, flower, and seed pods.

YARROW

Achillea millefolium L.

Perennial; hairy, with spreading rootstocks. Leaves are something like a carrot's. Flowers are white or yellow in dense, flat clusters at stem ends. Flowers are not normally found in turf.

4

Insect and Related Pests of Turfgrass

A. D. ALI, EXTENSION ENTOMOLOGIST, UNIVERSITY OF CALIFORNIA, RIVERSIDE

Most turfgrass injury from insect pests can be prevented by regular inspection of lawn areas and immediate remedial action. Insects may cause disorders such as die-back, stunting, and distortions of growth, as well as browning, yellowing, and bleaching of leaves. By detecting such symptoms early, you may be able to prevent the rapid buildup of insect populations possible under favorable conditions. Some pests feed only at night, and, unless you make a special effort to look for them, they may go undetected for a long time. The table at the end of this chapter will help you identify and treat a variety of insect and related pests.

Good control depends on correct identification of the pest and some knowledge of its behavior and biology—as well as on environmental factors such as temperature, moisture, soil type, and location that affect population buildup. Some pests thrive where it is warm and dry; others prefer wet or moist conditions. Some pests are influenced by cycles of drought. The frit fly, for example, is more prevalent during drought periods. Some locations in turf may be more susceptible to infestation than others. Often, pests are first found in isolated spots rather than distributed evenly throughout the turf. Edge effects are common, with the first infestation occurring near the turf border. Additionally, conditions of shade or sun, slope, and soil type can influence pest establishment.

INSECT DETECTION

Cutworms, sod webworms, lucerne moth larvae, skipper larvae, and vegetable weevil larvae can be detected with the pyrethrum test. Mix 1 tablespoonful of a commercial garden insecticide containing 1 to 2 percent pyrethrum in 1 gallon of water. Mark off 1 square yard of lawn area, including some damaged and some undamaged grass, and apply the entire gallon of mixture as evenly as possible to that area with a sprinkling can. Pyrethrum is irritating to many insects, and within 10 minutes they will come to the surface where they may be seen. If no pyrethrum is available, similar results can be obtained by flooding the area with water. Place the end of a garden hose on the area to be checked and let the water run for 5 to 10 minutes. The insects present will soon surface. Several areas of the turf should be checked to determine the extent of an infestation.

Neither white grubs nor billbug larvae will respond to the pyrethrum test or to flooding. Instead, carefully dig around the roots of grass. If infestations are heavy, grass roots will be eaten away, and the grass can be rolled back like a carpet. If more than 5 cutworms, 10 skipper larvae, or 15 sod webworms appear on average per square yard, or more than one white grub or billbug larva is found per square foot of established lawn, control measures should be taken.

In large lawn areas such as parks, golf courses, and cemeteries, perform pyrethrum tests or other pest inspections at several locations in order to determine the extent of an infestation. In some cases, the problem may be localized, and treatment of the entire turf area may be unnecessary.

No control measures will be necessary for many of the insects you will find, since they are not pests; the accurate identification of insects in lawns is essential. Furthermore, different pests require different chemical treatments and application methods for effective control.

CULTURAL PRACTICES

Some of the symptoms discussed above may be caused by disease, unfavorable soil conditions, or poor cultural practices. Investigate these possibilities before taking any pest control measures. In fact, the first line of defense against turfgrass pests is a program combining good cultural practices (watering, mowing, fertilization, aeration, and thatch control) and proper selection of the right turfgrass for the specific area and use. Poorly kept turf shows pest injury sooner and recovers more slowly than vigorous, well-kept turf.

Thatch, if allowed to accumulate and develop a thick layer of organic matter on the soil surface, can greatly influence the amount of damage an insect pest causes. Initially, thatch provides an ideal habitat and hiding place for insects, and their populations can increase. Also, insecticides become bound to the organic matter in thatch and cannot get into the soil efficiently. Thatch buildup results in poor water penetration and aeration, weakening the grass and making it more susceptible to insect damage. Even though damage can occur where thatch is not a factor, many cases of severe damage are associated with thatch buildup.

GUIDE TO TURFGRASS PEST CONTROL

(to be used with *Turfgrass Pests,* Publication 4053)

JUNE 1988

Cooperative Extension **University of California**
Division of Agriculture and Natural Resources

LEAFLET 2209

For information about ordering this publication
or about ordering Publication 4053,
Turfgrass Pests, write to:

Publications
Division of Agriculture and Natural Resources
University of California
6701 San Pablo Avenue
Oakland, California 94608-1239

or telephone (415) 642-2431

Leaflet 2209

Printed in the United States of America.

This leaflet supplies information concerning chemicals that can be recommended for use in controlling pests of turfgrass. It cannot be used properly without *Turfgrass Pests,* Publication 4053, also published by the University of California, Division of Agriculture and Natural Resources. To obtain that publication, see the ordering instructions above.

Because of frequent changes in regulations, as well as new discoveries, information regarding chemicals is often short-lived. What can be recommended today may be declared illegal next week—or a new and better chemical may be discovered the week after. This leaflet, therefore, will be revised and reissued as often as necessary to keep it current. So—BEFORE APPLYING ANY CHEMICAL LISTED HERE—note the date on the cover of this leaflet. If it is not current, please check with your local University of California county farm advisor, or write to Publications at the address above. **MOST IMPORTANT,** check the pesticide label for the specific use before application.

GUIDE TO TURFGRASS PEST CONTROL

(to be used with *Turfgrass Pests,* Publication 4053)

THE AUTHORS

Information and tables herein were prepared on—

- **weed control** by Clyde L. Elmore and W. B. McHenry, Extension Weed Scientists, Davis; and David W. Cudney, Extension Weed Scientist, Riverside
- **disease control** by Arthur H. McCain, Extension Plant Pathologist, Berkeley; and Robert M. Endo, Professor of Plant Pathology, and Howard D. Ohr, Extension Plant Pathologist, Riverside
- **insect control** by A. D. Ali, Extension Entomologist, and Timothy D. Paine, Assistant Professor of Entomology, Riverside
- **nematode control** by J. D. Radewald, Extension Nematologist, Riverside; and Becky Westerdahl, Extension Nematologist, Davis
- **rodent control** by Terrell P. Salmon, Extension Wildlife Specialist, Davis

HOW TO USE THIS LEAFLET

The companion publication *Turfgrass Pests* will help you identify your pest, whether it be an insect, disease, weed, or rodent. In most cases, we will suggest cultural practices or other nonchemical means to avoid or control problems. More often than not, cultural practices are safer and cheaper than pesticide applications. This is true not only from a curative approach but—more importantly—from a preventive one as well. Proper turf variety selection, irrigation, fertilization, mowing height, and thatch removal will result in healthy, vigorous turf. The increased vigor allows the turf to withstand insect, disease, and nematode damage better and to recover more quickly. Healthy turf also can out-compete weeds and reduce the chances of their becoming established. The net result will be less pesticide use and less hazard of human exposure.

If chemical control is still called for, this leaflet lists one or more materials that can be recommended. In all cases:

1. Observe safety precautions. *Safe* handling and application are as important as *effective* application.

2. Follow manufacturer's recommendations as to dosage. Applying more can be dangerous, injurious to turf, and costly; applying less can result in poor control—or no control at all.

3. Keep in mind that you, the applicator, are responsible for any damage done to neighboring turf or to other plants as a result of your material's drift.

4. Find out if you need a use permit from the County Agricultural Commissioner. Certain pesticides or groups of pesticides are designated as restricted materials. Their possession and use are subject to special restrictions under regulations of the California Department of Food and Agriculture. These pesticides include insecticides, herbicides, fungicides, fumigants, rodenticides, and avicides. Materials that require a use permit are designated in this publication with a footnote.

5. Observe all posting or re-entry requirements for all pesticides.

Also, be warned—

Remember that certain pesticides are toxic to plants if used at the wrong stage of plant development or when temperatures are too high. Injury may also result from an excessive application or the wrong formulation, or from mixing incompatible materials. Inert ingredients, such as wetters, spreaders, emulsifiers, diluents, and solvents can cause plant injury. Formulations are often changed by manufacturers, so plant injury may occur even though no injury was noted in previous seasons.

WEED CONTROL

To use the tables in this section with the companion publication *Turfgrass Pests* (Publication 4053),

1. Identify the weed or weeds you want to control by looking at the color photographs of Common Turfgrass Weeds in *Turfgrass Pests*.
2. Determine the best time of year for control measures by consulting the Characteristics of Turfgrass Weeds chart in *Turfgrass Pests*.
3. Use the table in this leaflet to determine the best chemical to apply.
4. Consult the Herbicide Formulation table in *Turfgrass Pests* for suggestions on how to apply the chemical.
5. For information regarding measurements, see the Measurements, Calculations, and Preparing the Sprayer section in *Turfgrass Pests*.
6. IN ALL CASES, follow the suggestions in the Safe and Effective Use of Pesticide Chemicals section in *Turfgrass Pests*.

HERBICIDE SELECTION GUIDE

		Treatment material. For specific rates, Selective in: (type of turf)	
Weed	Nonselective spot treatment in established turf	Bentgrass (*Agrostis* sp.)	Bermudagrass (*Cynodon dactylon*)
Broadleaf annual			
Black medic *Medicago lupulina*	dicamba*	dicamba*† 2,4-D +mecoprop +dicamba*§	dicamba*† 2,4-D+mecoprop +dicamba*
Chickweed *Stellaria media*	dicamba*	dicamba*†	dicamba*† 2,4-D+mecoprop +dicamba* benefin benefin +oryzalin¶ pendimethalin¶ benefin +trifluralin¶
Clover, California bur *Medicago polymorpha* var. *vulgaris*	dicamba* mecoprop	dicamba*† mecoprop† 2,4-D +mecoprop +dicamba*§	dicamba*† mecoprop† 2,4-D +mecoprop +dicamba*§
Geranium, cutleaf *Geranium dissectum*	2,4-D amine*†	2,4-D amine*†#	2,4-D amine*† 2,4-D +mecoprop +dicamba*

* Restricted herbicide: permit required from County Agricultural Commissioner.

† When a single product is mentioned, you can assume that combinations containing that product will also give control.

‡ Either the weed is not a problem or no chemical is available for the weed in this situation.

timing, etc., see *HERBICIDE SPECIFICATIONS*

	Selective in: (type of turf)			
Kentucky bluegrass or perennial ryegrass *(Poa pratensis or Lolium perenne)*	**Fescue** *(Festuca arundinacea)*	**Zoysia**	**Dichondra** *(Dichondra micrantha)*	
dicamba*† 2,4-D + mecoprop + dicamba* triclopyr	dicamba*† 2,4-D + mecoprop + dicamba* triclopyr	——‡	——‡	
dicamba*† 2,4-D + mecoprop + dicamba* triclopyr benefin¶ benefin + trifluralin¶ pendimethalin¶	dicamba*† 2,4-D + mecoprop + dicamba* triclopyr benefin¶ benefin + trifluralin¶	——‡	diphenamid¶ napropamide¶ benefin + oryzalin¶ benefin + trifluralin¶	
dicamba*† mecoprop† 2,4-D + mecoprop + dicamba*§	dicamba*† mecoprop† 2,4-D + mecoprop + dicamba*§	——‡	diphenamid¶ napropamide¶	
2,4-D amine*† 2,4-D + mecoprop + dicamba*	2,4-D amine*† 2,4-D + mecoprop + dicamba*	——‡	——‡	

Continued on next page

§ Use bentgrass formulation only.
¶ Must be applied before weed emergence.
\# 2,4-D can injure bentgrass. Use low rates or special formulations.
** Seedling applications only.

HERBICIDE SELECTION GUIDE—*Continued*

| | | Treatment material. For specific rates, | |
| | | Selective in: (type of turf) | |
Weed	Nonselective spot treatment in established turf	Bentgrass *(Agrostis* sp.*)*	Bermudagrass *(Cynodon dactylon)*
Broadleaf annual—*Continued*			
Knotweed, prostrate *Polygonum aviculare*	dicamba*	dicamba* mecoprop 2,4-D amine*#	dicamba* mecoprop 2,4-D ester* 2,4-D + mecoprop + dicamba* benefin + oryzalin¶ pendimethalin¶ benefin¶ benefin + oxadiazon¶ benefin + trifluralin¶
Lettuce, prickly *Lactuca scariola*	2,4-D amine*	2,4-D amine*#	2,4-D amine* benefin + oxadiazon¶
Mallow, little **(cheeseweed)** *Malva parviflora* (biennial)	2,4-D ester*	2,4-D amine*#	2,4-D ester* 2,4-D + mecoprop + dicamba* benefin + oxadiazon¶
Oxtongue, bristly *Picris echioides* (biennial)	2,4-D amine*	2,4-D amine*#	2,4-D amine* 2,4-D + mecoprop + dicamba*
Pearlwort, birdseye *Sagina procumbens*	mecoprop	dicamba*† mecoprop†	dicamba*† mecoprop† pendimethalin¶ benefin + oryzalin¶

* Restricted herbicide: permit required from County Agricultural Commissioner.

† When a single product is mentioned, you can assume that combinations containing that product will also give control.

‡ Either the weed is not a problem or no chemical is available for the weed in this situation.

timing, etc., see *HERBICIDE SPECIFICATIONS*

Selective in: (type of turf)			
Kentucky bluegrass or perennial ryegrass *(Poa pratensis or Lolium perenne)*	**Fescue** *(Festuca arundinacea)*	**Zoysia**	**Dichondra** *(Dichondra micrantha)*
dicamba* mecoprop 2,4-D ester* 2,4-D + mecoprop + dicamba* pendimethalin¶ benefin¶ benefin + trifluralin¶ benefin + oxadiazon (bluegrass only)¶	dicamba* mecoprop 2,4-D ester* 2,4-D + mecoprop + dicamba* pendimethalin¶ benefin¶ benefin + trifluralin¶	benefin + oryzalin¶ pendimethalin¶ benefin + trifluralin¶	napropamide¶
2,4-D amine* benefin + oxadiazon (bluegrass only)¶	2,4-D amine*	—‡	—‡
2,4-D ester* 2,4-D + mecoprop + dicamba* benefin + oxadiazon (bluegrass only)¶	2,4-D ester* 2,4-D + mecoprop + dicamba*	—‡	—‡
2,4-D amine* 2,4-D + mecoprop + dicamba*	2,4-D amine* 2,4-D + mecoprop + dicamba*	—‡	—‡
dicamba*† mecoprop† pendimethalin¶ benefin + trifluralin¶	dicamba*† mecoprop† pendimethalin¶ benefin + trifluralin¶	benefin + oryzalin¶ benefin + trifluralin¶	napropamide¶

Continued on next page

§ Use bentgrass formulation only.
¶ Must be applied before weed emergence.
# 2,4-D can injure bentgrass. Use low rates or special formulations.
** Seedling applications only.

HERBICIDE SELECTION GUIDE—*Continued*

		Treatment material. For specific rates, Selective in: (type of turf)	
Weed	Nonselective spot treatment in established turf	Bentgrass *(Agrostis* sp.*)*	Bermudagrass *(Cynodon dactylon)*
Broadleaf annual—Continued			
Pigweed, redroot *Amaranthus retroflexus*	2,4-D amine*	2,4-D amine*†# dicamba*†	2,4-D amine*† dicamba*† pendimethalin¶ benefin¶ benefin +oxadiazon¶ benefin +oryzalin¶
Pimpernel, scarlet *Anagallis arvensis*	2,4-D amine*	2,4-D amine*#	2,4-D amine*† pendimethalin¶ benefin +oryzalin¶ benefin +oxadiazon¶
Purslane, common *Portulaca oleracea*	2,4-D amine*	2,4-D amine*#	DCPA¶ 2,4-D amine*† pendimethalin¶ benefin¶ benefin +oryzalin¶ benefin +trifluralin¶ benefin +oxadiazon¶
Soliva (spurweed) *Soliva sessilis*	2,4-D ester*#	2,4-D amine*#	2,4-D amine* bromoxynil
Speedwell, birdseye *Veronica persica*	——‡	——‡	benefin¶ 2,4-D+mecoprop +dicamba*

* Restricted herbicide: permit required from County Agricultural Commissioner.

† When a single product is mentioned, you can assume that combinations containing that product will also give control.

‡ Either the weed is not a problem or no chemical is available for the weed in this situation.

timing, etc., see *HERBICIDE SPECIFICATIONS*

Selective in: (type of turf)			
Kentucky bluegrass or perennial ryegrass (*Poa pratensis* or *Lolium perenne*)	**Fescue** (*Festuca arundinacea*)	**Zoysia**	**Dichondra** (*Dichondra micrantha*)
2,4-D amine*† dicamba*† pendimethalin¶ benefin¶ benefin + oxadiazon (bluegrass only)¶ benefin + trifluralin¶	2,4-D amine*† dicamba*† pendimethalin¶ benefin + trifluralin¶	benefin + oryzalin¶ benefin + trifluralin¶	napropamide¶
2,4-D amine*† pendimethalin¶ benefin + trifluralin¶ benefin + oxadiazon (bluegrass only)¶	2,4-D amine*† pendimethalin¶ benefin + trifluralin¶	benefin + oryzalin¶	napropamide¶
DCPA¶ 2,4-D amine*† pendimethalin¶ benefin¶ benefin + trifluralin¶ benefin + oxadiazon (bluegrass only)¶	DCPA¶ 2,4-D amine*† pendimethalin¶ benefin¶ benefin + trifluralin¶	benefin + oryzalin¶ benefin¶	napropamide¶
2,4-D ester* bromoxynil	2,4-D ester* bromoxynil	bromoxynil + MSMA	——‡
benefin¶ 2,4-D + mecoprop ı dicamba*	——‡	——‡	——‡

Continued on next page

§ Use bentgrass formulation only.
¶ Must be applied before weed emergence.
# 2,4-D can injure bentgrass. Use low rates or special formulations.
**Seedling applications only.

HERBICIDE SELECTION GUIDE—*Continued*

Weed	Nonselective spot treatment in established turf	Treatment material. For specific rates, Selective in: (type of turf)	
		Bentgrass *(Agrostis* sp.*)*	**Bermudagrass** *(Cynodon dactylon)*
Broadleaf annual—Continued			
Spurge, spotted *Euphorbia maculata*	glyphosate	—‡	bromoxynil DCPA¶ pendimethalin¶ benefin +oryzalin¶
Broadleaf perennial			
Bindweed, field *Convolvulus arvensis*	2,4-D amine*	2,4-D amine*#	2,4-D+mecoprop +dicamba* benefin +oryzalin¶ pendimethalin¶**
Catsear, spotted *Hypochoeris radicata*	2,4-D amine*	2,4-D amine*# 2,4-D+mecoprop +dicamba*§	2,4-D amine* 2,4-D+mecoprop +dicamba*
Chickweed, mouseear *Cerastium vulgatum*	mecoprop	dicamba* mecoprop 2,4-D +mecoprop +dicamba*	dicamba* mecoprop 2,4-D+mecoprop +dicamba*
Clover, white *Trifolium repens*	mecoprop dicamba*	dicamba* mecoprop 2,4-D +mecoprop +dicamba*	dicamba* mecoprop 2,4-D+mecoprop +dicamba*
Daisy, English *Bellis perennis*	dicamba*	dicamba* 2,4-D +mecoprop +dicamba*§	dicamba* dicamba+2,4-D* 2,4-D +mecoprop +dicamba*

* Restricted herbicide: permit required from County Agricultural Commissioner.

† When a single product is mentioned, you can assume that combinations containing that product will also give control.

‡ Either the weed is not a problem or no chemical is available for the weed in this situation.

timing, etc., see *HERBICIDE SPECIFICATIONS*

Selective in: (type of turf)			
Kentucky bluegrass or perennial ryegrass *(Poa pratensis or Lolium perenne)*	**Fescue** *(Festuca arundinacea)*	**Zoysia**	**Dichondra** *(Dichondra micrantha)*
bromoxynil DCPA¶ pendimethalin¶ benefin +trifluralin¶	DCPA¶ pendimethalin¶ benefin +trifluralin¶	benefin +oryzalin¶ pendimethalin¶	——‡
2,4-D+mecoprop +dicamba* pendimethalin¶** benefin +trifluralin¶**	2,4-D+mecoprop +dicamba* pendimethalin¶** benefin +trifluralin¶**	pendimethalin¶** benefin+oryzalin¶**	——‡
2,4-D amine* 2,4-D+mecoprop +dicamba*	2,4-D amine* 2,4-D+mecoprop +dicamba*	——‡	——‡
dicamba* mecoprop 2,4-D+mecoprop +dicamba*	dicamba* mecoprop 2,4-D+mecoprop +dicamba*	——‡	——‡
dicamba* mecoprop 2,4-D+mecoprop +dicamba* triclopyr	dicamba* mecoprop 2,4-D+mecoprop +dicamba* triclopyr	——‡	monuron neburon
dicamba* dicamba+2,4-D* 2,4-D +mecoprop +dicamba*	dicamba* dicamba+2,4-D* 2,4-D +mecoprop +dicamba*	——‡	——‡

Continued on next page

§ Use bentgrass formulation only.
¶ Must be applied before weed emergence.
# 2,4-D can injure bentgrass. Use low rates or special formulations.
**Seedling applications only.

HERBICIDE SELECTION GUIDE—*Continued*

Treatment material. For specific rates,

Selective in: (type of turf)

Weed	Nonselective spot treatment in established turf	Bentgrass *(Agrostis* sp.*)*	Bermudagrass *(Cynodon dactylon)*
Broadleaf perennials—Continued			
Dandelion *Taraxacum officinale*	2,4-D amine*	2,4-D amine*# 2,4-D +mecoprop +dicamba*§	2,4-D amine* 2,4-D+mecoprop +dicamba*
Dichondra *Dichondra micrantha*	—‡	2,4-D amine*#	2,4-D ester*
Dock, curly *Rumex crispus*	dicamba* 2,4-D amine* triclopyr	dicamba* 2,4-D amine*# 2,4-D +mecoprop +dicamba*§	dicamba* 2,4-D amine* 2,4-D+mecoprop* +dicamba*
Healall (selfheal) *Prunella vulgaris*	—‡	dicamba+2,4-D* 2,4-D +mecoprop +dicamba*§	dicamba+2,4-D* 2,4-D +mecoprop +dicamba*§
Plantain, broadleaf *Plantago major*	2,4-D amine*	2,4-D amine*# 2,4-D +mecoprop +dicamba*§	2,4-D amine* 2,4-D+mecoprop +dicamba*
Plantain, buckhorn *Plantago lanceolata*	2,4-D amine*	2,4-D amine*# 2,4-D +mecoprop +dicamba*§	2,4-D amine* 1,4-DD+mecoprop +dicamba*
Sorrel, red *Rumex acetosella*	dicamba*	dicamba*	dicamba* 2,4-D+mecoprop +dicamba*
Woodsorrel, creeping *Oxalis corniculata*	triclopyr	—‡	2,4-D+mecoprop +dicamba +MSMA* pendimethalin¶** benefin +oryzalin¶**

* Restricted herbicide: permit required from County Agricultural Commissioner.

† When a single product is mentioned, you can assume that combinations containing that product will also give control.

‡ Either the weed is not a problem or no chemical is available for the weed in this situation.

timing, etc., see *HERBICIDE SPECIFICATIONS*

Selective in: (type of turf)			
Kentucky bluegrass or perennial ryegrass (*Poa pratensis* or *Lolium perenne*)	**Fescue** (*Festuca arundinacea*)	**Zoysia**	**Dichondra** (*Dichondra micrantha*)
2,4-D amine* 2,4-D + mecoprop + dicamba*	2,4-D amine* 2,4-D + mecoprop + dicamba*	—‡	—‡
2,4-D ester*	2,4-D ester*	—‡	—‡
dicamba* 2,4-D amine* 2,4-D + mecoprop + dicamba* triclopyr	dicamba* 2,4-D amine* 2,4-D + mecoprop + dicamba* triclopyr		
dicamba + 2,4-D* 2,4-D + mecoprop + dicamba*	dicamba + 2,4-D*	—‡	—‡
2,4-D amine* 2,4-D + mecoprop + dicamba*	2,4-D amine* 2,4-D + mecoprop + dicamba*	—‡	—‡
2,4-D amine* 2,4-D + mecoprop + dicamba*	2,4-D amine* 2,4-D + mecoprop + dicamba*	—‡	—‡
dicamba* 2,4-D + mecoprop + dicamba*	dicamba* 2,4-D + mecoprop + dicamba*	—‡	—‡
triclopyr pendimethalin¶** benefin + trifluralin¶**	triclopyr pendimethalin¶** benefin + trifluralin¶**	pendimethalin¶** benefin + oryzalin¶**	—‡

Continued on next page

§ Use bentgrass formulation only.
¶ Must be applied before weed emergence.
# 2,4-D can injure bentgrass. Use low rates or special formulations.
**Seedling applications only.

HERBICIDE SELECTION GUIDE—*Continued*

Weed	Nonselective spot treatment in established turf	Treatment material. For specific rates, Selective in: (type of turf)	
		Bentgrass *(Agrostis* sp.)	Bermudagrass *(Cynodon dactylon)*
Broadleaf perennial—Continued			
Yarrow, common *Achilea millefolium*	2,4-D ester*	——‡	2,4-D ester* dicamba* 2,4-D+mecoprop +dicamba*
Narrowleaf annual			
Barnyardgrass *Echinochloa crusgalli*	glyphosate	bensulide¶	bensulide¶ DCPA¶ benefin¶ benefin +oryzalin¶ pendimethalin¶ benefin +oxadiazon¶
Bluegrass, annual *Poa annua*	glyphosate	bensulide¶	bensulide¶ benefin¶ oxadiazon¶ pronamide¶ pendimethalin¶ benefin +trifluralin¶ benefin +oxadiazon¶
Foxtail, yellow *Setaria glauca*	glyphosate	——‡	bensulide¶ DCPA¶ benefin¶ pronamide¶ benefin +oryzalin¶ benefin +trifluralin¶

* Restricted herbicide: permit required from County Agricultural Commissioner.

† When a single product is mentioned, you can assume that combinations containing that product will also give control.

‡ Either the weed is not a problem or no chemical is available for the weed in this situation.

timing, etc., see *HERBICIDE SPECIFICATIONS*

	Selective in: (type of turf)		
Kentucky bluegrass or perennial ryegrass *(Poa pratensis or Lolium perenne)*	**Fescue** *(Festuca arundinacea)*	**Zoysia**	**Dichondra** *(Dichondra micrantha)*
2,4-D ester* dicamba* 2,4-D+mecoprop +dicamba*	2,4-D ester* dicamba* 2,4-D+mecoprop +dicamba*	—‡	—‡
bensulide DCPA¶ benefin¶ pendimethalin¶ benefin +oxadiazon (bluegrass only)¶ benefin +trifluralin¶	bensulide¶ DCPA¶ benefin¶ pendimethalin¶ benefin +trifluralin¶	bensulide¶ benefin +oryzalin¶ pendimethalin¶	napropamide¶ bensulide¶ diphenamid¶ fluazifop
bensulide¶ benefin¶ oxadiazon¶ pendimethalin¶ benefin +trifluralin¶ benefin +oxadiazon (bluegrass only)¶	bensulide¶ benefin¶ pendimethalin¶ benefin +trifluralin¶	bensulide¶ pendimethalin¶ benefin +oryzalin¶	bensulide¶ diphenamid¶ napropamide¶
bensulide¶ DCPA¶ benefin¶ pendimethalin¶ benefin +trifluralin¶	bensulide¶ DCPA¶ benefin¶ pendimethalin¶ benefin +trifluralin¶	bensulide¶ benefin +oryzalin¶ pendimethalin¶	napropamide¶

Continued on next page

§ Use bentgrass formulation only.
¶ Must be applied before weed emergence.
\# 2,4-D can injure bentgrass. Use low rates or special formulations.
** Seedling applications only.

HERBICIDE SELECTION GUIDE—*Continued*

Weed	Nonselective spot treatment in established turf	Treatment material. For specific rates, Selective in: (type of turf)	
		Bentgrass (*Agrostis* sp.)	Bermudagrass (*Cynodon dactylon*)
Narrowleaf annual—Continued			
Crabgrass, large *Digitaria sanguinalis*	glyphosate¶	bensulide¶ DSMA¶	benefin¶ bensulide¶ DCPA¶ DSMA MSMA CMA benefin +oxadiazon¶ benefin +oryzalin¶ pendimethalin¶ benefin +trifluralin¶
Crabgrass, smooth *Digitaria ischaemun*	glyphosate	bensulide¶ DSMA	benefin¶ bensulide¶ DCPA¶ DSMA CMA MSMA benefin +oxadiazon¶ benefin +oryzalin¶ pendimethalin¶ benefin +trifluralin¶
Goosegrass *Eleusine indica*	glyphosate	bensulide¶	bensulide¶ pendimethalin¶ benefin +oxadiazon¶ benefin +oryzalin¶ benefin +trifluralin¶

* Restricted herbicide: permit required from County Agricultural Commissioner.

† When a single product is mentioned, you can assume that combinations containing that product will also give control.

‡ Either the weed is not a problem or no chemical is available for the weed in this situation.

timing, etc., see *HERBICIDE SPECIFICATIONS*

Selective in: (type of turf)

Kentucky bluegrass or perennial ryegrass (*Poa pratensis* or *Lolium perenne*)	Fescue (*Festuca arundinacea*)	Zoysia	Dichondra (*Dichondra micrantha*)
benefin¶ bensulide¶ DCPA¶ DSMA MSMA CMA benefin +trifluralin¶ benefin +oxadiazon (bluegrass only)¶ pendimethalin¶	benefin¶ bensulide¶ DCPA¶ DSMA MSMA CMA benefin +trifluralin¶ pendimethalin¶	pendimethalin¶ MSMA¶ benefin +oryzalin¶	bensulide¶ diphenamid¶ napropamide¶ fluazifop¶
benefin¶ bensulide¶ DCPA¶ DSMA CMA MSMA benefin +trifluralin¶ benefin +oxadiazon¶ pendimethalin¶	benefin¶ bensulide¶ DCPA¶ DSMA CMA MSMA benefin +trifluralin¶ pendimethalin (bluegrass only)¶	MSMA pendimethalin¶ benefin +oryzalin¶	bensulide¶ diphenamid¶ napropamide¶ fluazifop
bensulide¶ pendimethalin¶ benefin +oxadiazon (bluegrass only)¶ benefin +trifluralin¶	bensulide pendimethalin¶ benefin +trifluralin¶	pendimethalin¶ benefin +oryzalin¶	bensulide¶ napropamide¶ fluazifop

Continued on next page

§ Use bentgrass formulation only.
¶ Must be applied before weed emergence.
\# 2,4-D can injure bentgrass. Use low rates or special formulations.
**Seedling applications only.

HERBICIDE SELECTION GUIDE—*Continued*

		Treatment material. For specific rates,	
		Selective in: (type of turf)	
Weed	Nonselective spot treatment in established turf	Bentgrass (*Agrostis* sp.)	Bermudagrass (*Cynodon dactylon*)
Narrowleaf annual—Continued			
Ryegrass, Italian *Lolium multiflorum*	glyphosate	—‡	pronamide¶ pendimethalin¶ benefin + oryzalin¶ benefin + oxadiazon¶ benefin + trifluralin¶
Narrowleaf perennial			
Bentgrass *Agrostis* sp.	dazomet metham glyphosate	—‡	2,4-D ester*
Bermudagrass *Cynodon dactylon*	dazomet metham glyphosate	—‡	—‡
Bluegrass, Kentucky *Poa pratensis*	dazomet metham glyphosate	—‡	—‡
Dallisgrass *Paspalum dilatatum*	dazomet metham glyphosate	DSMA	DSMA MSMA pendimethalin¶** benefin + oryzalin¶**
Fescue, tall *Festuca arundinacea*	dazomet metham glyphosate	—‡	—‡
Kikuyugrass *Pennisetum clandestinum*	dazomet metham glyphosate	—‡	—‡

* Restricted herbicide: permit required from County Agricultural Commissioner.

† When a single product is mentioned, you can assume that combinations containing that product will also give control.

‡ Either the weed is not a problem or no chemical is available for the weed in this situation.

timing, etc., see *HERBICIDE SPECIFICATIONS*

Selective in: (type of turf)

Kentucky bluegrass or perennial ryegrass *(Poa pratensis or Lolium perenne)*	Fescue *(Festuca arundinacea)*	Zoysia	Dichondra *(Dichondra micrantha)*
pendimethalin¶ benefin +oxadiazon (bluegrass only)¶ benefin +trifluralin¶	pendimethalin¶ benefin +trifluralin¶	benefin +oryzalin¶	fluazifop napropamide¶
——‡	——‡	——‡	fluazifop
siduron	siduron	——‡	fluazifop
——‡	——‡	——‡	diphenamid¶
DSMA MSMA pendimethalin¶** benefin +trifluralin¶**	DSMA MSMA pendimethalin¶** benefin +trifluralin¶**	benefin +oryzalin¶** pendimethalin¶**	fluazifop napropamide¶
——‡	——‡	——‡	fluazifop
siduron	siduron	——‡	fluazifop

Continued on next page

§ Use bentgrass formulation only.
¶ Must be applied before weed emergence.
# 2,4-D can injure bentgrass. Use low rates or special formulations.
**Seedling applications only.

HERBICIDE SELECTION GUIDE—*Continued*

		Treatment material. For specific rates,	
		Selective in: (type of turf)	
Weed	Nonselective spot treatment in established turf	Bentgrass (*Agrostis* sp.)	Bermudagrass (*Cynodon dactylon*)
Narrowleaf perennial—Continued			
Nutsedge, yellow *Cyperus esculentus*	glyphosate bentazon	DSMA	DSMA MSMA bentazon
St. Augustinegrass *Stenotaphrum secundatum*	dalapon dazomet metham glyphosate	2,4-D amine*#	2,4-D ester*
Velvetgrass, German *Holcus mollis*	metham glyphosate dazomet	——‡	——‡

* Restricted herbicide: permit required from County Agricultural Commissioner.

† When a single product is mentioned, you can assume that combinations containing that product will also give control.

‡ Either the weed is not a problem or no chemical is available for the weed in this situation.

timing, etc., see *HERBICIDE SPECIFICATIONS*

Selective in: (type of turf)

Kentucky bluegrass or perennial ryegrass *(Poa pratensis or Lolium perenne)*	**Fescue** *(Festuca arundinacea)*	**Zoysia**	**Dichondra** *(Dichondra micrantha)*
DSMA MSMA bentazon	DSMA MSMA bentazon	DSMA MSMA	—‡
2,4-D ester*	2,4-D ester*	—‡	—‡
—‡	—‡	—‡	fluazifop

§ Use bentgrass formulation only.
¶ Must be applied before weed emergence.
\# 2,4-D can injure bentgrass. Use low rates or special formulations.
**Seedling applications only.

HERBICIDE SPECIFICATIONS

Trade names given here in parentheses next to the chemical names do not constitute an endorsement by the University of California, nor do they indicate criticism of any similar products not included in the listing.

Where a single herbicide is listed for a weed species, it is the active compound for control. The same herbicide may be found in combination or may be combined by the applicator with other products, unless the label specifically prohibits mixing the two products. Combinations generally broaden the range of weed species controlled.

Read and follow the label instructions as to the herbicide application rate and the safe handling and application of the herbicide.

NOTE: ai/A = active ingredient per acre; ae = acid equivalent

benefin (Balan)

Formulation: Granules or with fertilizer.
Use: Selective, soil-applied, preemergent.
Rate: 3 lb ai/A.
Remarks: a) For crabgrass: Apply 2 to 3 weeks before initial germination (January for Los Angeles, Basin, early to mid-February for Central Valley and central coast, mid-February to March 1 for northern California and north coastal areas). Sprinkle-irrigate after application to wash herbicide off leaves and into the soil.
b) For annual bluegrass: Apply 2 to 3 weeks before initial germination (August-September). Sprinkle-irrigate after application to wash herbicide off leaves and into the soil.
c) For speedwell: Apply preemergence in January.
d) Often combined with other preemergence herbicides for longer residual.
NOTE: Do not apply to bentgrass greens.

benefin plus oryzalin (XL)

Formulation: Granules or with fertilizer.
Use: Selective, soil-applied, preemergent.
Rate: 2 to 3 lb ai/A.
Remarks: For use on warm season grasses only. Apply on established turf prior to germination of annual weeds. Do not aerate or verticut after application. Do not overseed with grasses for 12 to 16 weeks after application. Do not use on bluegrass, ryegrass, or fescue turf.

benefin plus trifluralin (Team)

Formulation: Granules or with fertilizer.
Use: Selective, soil-applied, preemergent.
Rate: *For cool season species:* 1.5 to 2 lb ai/A. *For warm season species:* 2 to 3 lb ai/A.
Remarks: Apply on established turf in the spring 1 to 2 weeks before expected germination of summer annuals (crabgrass, goosegrass, foxtail, or barnyardgrass). For annual bluegrass control: apply in the late summer or early fall before germination. A second application can be applied 10 to 12 weeks after the first in the southern part of the state to control late-germinating weeds. Do not overseed grasses for 12 to 16 weeks after application.

bensulide (Betasan, Scotts Halts, Scotts Super Halts Plus, Super Pax Crabgrass Control with Betasan, Presan)

Formulation: Emulsifiable liquid, granular, with fertilizer.
Use: Selective, soil-applied, preemergent.
Rate: 7.5 to 10 lb ai/A.
Remarks: Safest preemergence control material in bentgrass. For crabgrass: Apply 2 to 3 weeks before initial germination (January for Los Angeles Basin and south coast area, early to mid-February for Central Valley and central coast, mid-February to March 1 for northern California and north coastal areas).
NOTE: Crabgrass may germinate and become established in turf in late summer if lower rates are used. Good management will allow use of lower rates.
Rate: 7.5 lb ai/A in fall and 7.5 lb ai/A in midwinter (Jan.-Feb.).
Remarks: For annual bluegrass: Apply in early fall before annual bluegrass germinates. Normally mid-August to mid-September.
NOTE: Exclude children and pets during application and until treated area has been thoroughly sprinkler-irrigated.

bentazon (Basagran)

Formulation: 4 lb ai/gal emulsifiable concentrate.
Use: Selective, postemergent.
Rate: 1 to 2 lb ai/A in 40 gal water/A.
Remarks: For yellow nutsedge in grass turf. The nutsedge should be growing vigorously with good soil moisture. If control is not as desired, apply a second treatment after 10 to 14 days. Do not apply more than 3 lb per season. For optimum control, do not mow 3 to 5 days before or after application. Basagran can be mixed with 2,4-D (Basagran 1 lb + 2,4-D 1 lb) for nutsedge and other broadleaf control. Do not use on newly seeded or sprigged turf.

bromoxynil (Buctril, Buctril 4EC, ME4 Brominal)

Formulation: 2 lb/gal emulsifiable concentrate; 4 lb/gal emulsifiable concentrate (Buctril 4EC, ME4 Brominal).
Use: Selective, foliage-applied, contact.
Rate: 0.375 to 0.5 lb ai/A at 3- to 4-leaf stage or up to 6-inch weed height, or on rosette plants before they exceed 1½ inches in diameter; 1 lb ai/A for seedling spurweed, spiny fruited crowfoot, black medic, and hop clover; 2 lb ai/A for seedling prostrate spurge.
Remarks: Particularly beneficial at 0.375 to 0.5 lb ai/A on seedling grasses aged 3 weeks or older to control broadleaf weeds. Use lower rates on small weeds and higher rates on large weeds. Apply in at least 20 gallons of water per acre. May be tank-mixed with other broadleaf materials such as 2,4-D and 2,4DP, MCPP, dicamba, MSMA, or DSMA, or combinations of these materials, depending upon the weed species present.

cacodylic acid (Scotts Spot Grass & Weed Control, Liquid Edger, Best Lawn Edger & Trimmer, Germain's Fresh Start Grass & Weed Killer, Acme Weed-N-Grass Killer, Rad-E-Cate 25)

Formulation: Soluble liquid and soluble powder.
Use: Nonselective, foliage-applied, postemergent, contact.
Rate: 6 to 8 lb ai+1 qt surfactant/A.
Remarks: Apply principally for the control of annual weeds before cultivating and planting turf.

CAMA (Super Dal-E-Rad Calar, Ortho Crabgrass Killer, Formula II)

Formulation: Soluble liquid or soluble powder.
Use: Selective, foliage-applied, translocated.
Rate: 2.0 to 2.5 lb ai/A.
Remarks: Temperature and turf type determine degree of selectivity. Use low rate for young crabgrass in fine-leaved fescue turf—or when daily high temperature exceeds 85°F.
Use high rate for mature crabgrass in Kentucky bluegrass or bermudagrass turf or when temperature is 85°F or less.
NOTE: Apply uniformly over area regardless of distribution of crabgrass. Hesitating over more weedy spots may cause excessive rate and injure or kill turf. Repeat at 5- to 7-day intervals for total of 2 to 3 treatments for crabgrass. May temporarily discolor turf. May injure St. Augustine, fine fescue, and some bentgrass species.

dazomet (Basamid)

Formulation: 99% granular.
Use: Nonselective, soil-applied, fumigant, preplant.
Rate: 275 lb ai/A or 10 oz/100 sq ft.
Remarks: Apply in water; mix into soil 6 inches deep with a power tiller. Seed in 3 weeks if temperature is over 60°F and soil is moist but not wet. Wash into soil with sprinkler irrigation when spot treating. Effective principally on annual weeds.

DCPA (Acme Garden Weed Preventer Granules, Acme Garden Weed Preventer Spray, Best DCPA 5 Granules, Dacthal W-75 for Turf, Pax Crabgrass & Spurge Preventer Plus Fertilizer, Dacthal G-5)

Formulation: Wettable powder, granular, formulated with fertilizer.
Use: Selective, soil-applied, preemergent.
Rate: 10 lb ai/A.
Remarks: Apply 2 to 3 weeks before initial crabgrass germination (January for Los Angeles Basin and south coast area, early to mid-February for Central Valley and central coast area, mid-February to March 1 for northern California and north coast area).
NOTE: Do not use on bentgrass and dichondra. Exclude children and pets during application and until treated area has been thoroughly sprinkler-irrigated. Young crabgrass plants may become established in turf in late summer if lower rates are used. NOTE: Will not control crabgrass after germination. Apply for annual bluegrass control at end of August or beginning of September.

dicamba (Banvel 4-S)

Formulation: Soluble liquid.
Use: Selective, foliage-applied, translocated.
Rate: 1/4 lb/A in 40 gal water.
Remarks: Apply for chickweeds, clovers, English daisy, prostrate knotweed, pearlwort, red sorrel, curly dock. Do not apply more than two times per year. Also for spot spraying (applies to 4 lb ae/gal formulation).
NOTE: Do not exceed 1/2 lb ae/A per season. Active through the soil; do not use where roots of ornamental plants may extend into treated area. Spray on calm days to avoid spray drift onto susceptible crops or ornamentals. Often used at low rates in combination with 2,4-D and mecoprop. Nonselective on dichondra.

dicamba + 2,4-D* (Super D Weedone, Scotts Lawn Weed Control, Scotts Spot Dandelion Control, Super Pax Weed'N Feed, Provel Lawn and Turf Herbicide)

Formulation: Soluble liquid, granular formulated with fertilizer.
Use: Selective, foliage-applied, translocated.
Rate: Varies with specific product.
Remarks: For English daisy or other difficult to control broadleaf weeds where there is dandelion or plantain present.
NOTE: Do not exceed 1/4 lb ae/A of dicamba on bentgrass turf. Active through the soil; do not use where roots of ornamentals may extend into treated area. Spray on calm days to avoid spray drift onto susceptible crops or ornamentals. Nonselective on dichondra.

dicamba + 2,4-D + mecoprop* (Lilly/Miller Feed & Weed, Lilly/Miller Spot Weeder, Miller's Lawn & Turf Weed Bomb, Best Lawn Weed Killer, Acme Super Weed-No-More, Acme Super Chickweed Killer, Acme Southern Weed-No-More, Acme Weed-No-More Spot Weeder, Miller's Lawn & Turf Weedkiller)

Formulation: Emulsifiable concentrate.
Use: Selective, foliage-applied, translocated.
Low Rate: 0.075 to 0.11 lb dicamba/A + 0.18 to 1.1 lb 2,4-D/A + 0.55 to 0.75 lb mecoprop/A.
High rate: 0.08 to 0.12 lb dicamba/A + 0.64 to 0.96 lb 2,4-D/A + 0.32 to 0.48 lb mecoprop/A.
Remarks: For broad-spectrum control of broadleaf weeds. Use lower rates for bentgrass, hybrid bermudagrass and other sensitive turfgrasses. Nonselective on dichondra. Avoid applying to drought- and heat-stressed turf. Do not irrigate within 24 hours of application. Newly seeded turf should not be treated until after the second or third mowing. Bentgrass most sensitive turfgrass. Read label for further application directions.

diphenamid (Enide)

Formulation: Wettable powder or granular formulated with fertilizer.
Use: Selective, soil-applied, preemergent.
Rate: 10 lb ai/A.
Remarks: a) For dichondra turf only: Effective preemergence and early postemergence mainly for grass control.
b) For annual bluegrass: Apply in early fall and spring.
c) For crabgrass: January for Los Angeles Basin; February for Central Valley and central coast; March for northern California and north coast area.

*Restricted herbicide; permit required from County Agricultural Commissioner.

    d) For bermudagrass: Apply to suppress growth, does not eradicate established plants.
   NOTE: Will seriously injure or kill turfgrasses. Exclude children and pets during application and until treated area has been thoroughly sprinkler-irrigated.

DSMA (Chacon Crabgrass Control, Weedone Crabgrass Killer, Scotts Summer Crabgrass Control, DSMA Liquid)

Formulation: Soluble liquid or soluble powder.
Use: Selective, foliage-applied, translocated.
Low Rate: 3 lb ai/A in 175 to 200 gal water. Use lower rate on bentgrasses and fine-leaved fescues. Sufficient rate for young crabgrass, and with repeated monthly sprays for established dallisgrass and nutsedge. Use if daily temperatures exceed 80°F.
High Rate: 4 lb ai/A in 175 to 200 gal water. Use higher rate for mature crabgrass. Requires 2 to 3 resprays at 5- to 7-day intervals. Satisfactory rate for use in bermudagrass, and if temperatures are 80°F or lower in Kentucky bluegrass as well. Will yellow zoysia turf.
Remarks: Effective in controlling crabgrass, dallisgrass, and nutsedge. Temperature, soil moisture, and turf type determine degree of turf-selectivity. Avoid spraying under hot, droughty conditions. Bents, fine-leaved fescues, and dichondra are most sensitive; bermudagrass is most tolerant. Do not use on St. Augustine turf.

fluazifop (Ortho Grass-B-Gon)

Formulation: 0.5% liquid.
Use: Selective, postemergent.
Rate: Spray directly from bottle according to label instructions.
Remarks: For selective grass control in dichondra only. Will not control annual bluegrass. Apply when the grass is young and vigorous and has good soil moisture. Retreatments may be required for hard-to-kill weeds such as bermudagrass, dallisgrass, and kikuyugrass. Will not control nutsedge.

glyphosate (Roundup, Ortho Kleenup Systemic Grass and Weed Killer, Lilly/Miller Knock-Out Weed & Grass Killer)

Formulation: Soluble liquid.
Use: Nonselective, foliage-applied, translocated.
Rate: 1 to 5 lb ai/A in 20 to 40 gal water or 2 to 3 oz/gal/1,000 sq ft.
Remarks: Apply to rapidly growing weeds. *Annual weeds:* If shorter than 6 inches, apply 1 lb ai/A; if 6 inches or taller, apply 1.5 lb ai/A. Allow minimum of 3 days between application and renovation

or cultivation. *Perennial weeds:* Apply to vigorous but nearly mature weeds (bermudagrass: summer to fall; field bindweed, at full bloom). Apply 4 to 5 lb ai/A.

In a mowed turf grass area, *omit at least one mowing* before application. Delay verticutting, removing sod or tillage for at least 7 days after treatment. To maximize control allow the soil surface and root area to dry after verticutting or sod removal before replanting. When turf or ornamentals are to be planted, a followup preemergence program is required to control the seed of perennials.

mecoprop (Chipco Turf Herbicide MCPP)

Formulation: Soluble liquid.
Use: Selective, foliage-applied, translocated.
Rate: 1.5 lb ae/A + 1 qt surfactant per 100 gal spray. *For spot spraying:* Use same concentration per 100 gal or 3 to 4 tsp + 2 tsp surfactant/gal water.
Remarks: Clover, prostrate knotweed, pearlwort.
NOTE: Spray on calm days to avoid spray drift onto susceptible crops or ornamentals. Prostrate knotweed should be treated when young (2 to 4 inches in diameter). Nonselective on dichondra. (Rate for spot spraying applies only to formulations containing 2 or 2.5 lb ae/gal.)

metham (Vapam)

Formulation: Soluble liquid.
Use: Preplant or nonselective soil-applied fumigant.
Rate: 430 lb ai/A or 10 lb ai/1,000 sq ft.
Remarks: Apply in water on calm day; follow immediately with sprinkler irrigation to seal the soil surface or, preferably, cover with vapor-proof covering. Seed in 2 weeks on light sandy soils, in 3 to 4 weeks on heavier clay or mulch (organic) soils. Extend waiting period if temperature is below 60°F. Two applications usually required to eradicate bermudagrass or kikuyugrass. Roto-tilling before treatment will enhance control.

MSMA (Acme Crabgrass & Nutgrass Killer, Germain's Improved Crabgrass Killer, Bueno 6)

Formulation: Soluble liquid.
Use: Selective, foliage-applied, translocated.
Rate: 3 to 4 lb ai/A.

Remarks: Temperature and turf type determine degree of selectivity. Use lower rate for nutsedge control, on bentgrass, and on other turf types when daily temperature exceeds 85°F. Apply at monthly intervals for control of dallisgrass and nutsedge.
NOTE: Apply uniformly over area regardless of distribution of the weed. Hesitating with sprayer over weedier spots may cause excessive rate and injure or kill the turf. Repeated applications of high rates reduces kikuyugrass. Turf may be temporarily discolored. Injurious to St. Augustine grass, red fescue, dichondra, and zoysia.

napropamide (Devrinol)

Formulation: 50% wettable powder, 5% granular.
Use: Selective, preemergent.
Rate: 2 to 3 lb ai/A.
Remarks: Apply at seeding or on established dichondra. Principally for grass control, but will control some broadleaf weeds. Follow treatment with a minimum of 1 inch of water to wash material from the leaves and into the soil.

oxadiazon (Ronstar)

Formulation: 50% wettable powder, 2% granular.
Use: Selective, soil-applied, preemergent.
Rate: 2 to 4 lb ai/A.
Remarks: Wettable powder to be used only on dormant established bermudagrass, St. Augustine, or zoysiagrass turf. The granule formulation can be used safely on most other grass species except bentgrass. Do not use on dichondra. Some foliar injury can be observed if the granules are applied to wet foliage or the herbicide is not washed from the leaves after application. Has not been effective for control of prostrate spurge or creeping woodsorrel (Oxalis) in California. Do not use on newly seeded turf. Apply the wettable powder formulation at least 2 weeks before turf greens in the spring.

oxadiazon plus benefin (Regalstar)

Formulation: 1% oxadiazon and 0.5% benefin on a ureaformaldehyde fertilizer.
Use: Selective, soil-applied, preemergent.
Rate: 2 lb ai/A oxadiazon and 1 lb/A benefin.
Remarks: Use only on bermudagrass and bluegrass for the control of summer annual weeds (crabgrass and goosegrass). May be used on

newly sprigged bermudagrass after stolons have rooted and filled in the bare spaces. Apply in early spring before crabgrass germination. Apply to newly seeded bermudagrass and bluegrass after grass has been mowed at least twice. For commercial use only; not for use on home lawns.

pronamide (Kerb)

Formulation: 50% wettable powder.
Use: Selective, preemergent, postemergent.
Rate: 0.5 to 1.5 lb ai/A.
Remarks: Apply rates necessary to control the stage of growth of the annual bluegrass. Apply 0.5 to 1.0 lb preemergence or early postemergence; 0.75 to 1.0 lb postemergence, early tillering to heading; 1.0 to 2.0 lb seed-forming stage. Apply the low rate to light sandy soils and the high rate to loamy and clayey soils. Control is slow to be observed (4 to 6 weeks). Do not apply to areas to be overseeded within 90 days of treatment. Use only on bermudagrass. Irrigate (1/2 inch) within 2 days to get pronamide into the root zone.

siduron (Tupersan)

Formulation: 50% wettable powder.
Use: Selective, soil-applied, preemergent.
Rate: 2 to 6 lb ai/A—new spring plantings (when no more than 3 lb are used at planting, a second application of 2 to 3 lb should be made 4 weeks later); 8 to 12 lb ai/A—fall plantings or established turf. Apply the low rates on sandy soils and higher rates on loamy and clayey soils.
Remarks: Used for the selective control of seedlings of summer grasses (including crabgrass, bermudagrass, and dallisgrass). Since seeds of these grasses occur for more than 1 year, a preemergence program must be followed for 2 years following renovation of warm season turf. Do not allow contact with the sprayed area until material has dried. At least 1/2 inch of water is required to wash the material off of the turf and into the soil.

triclopyr (Turflon)

Formulation: 4 lb/gal emulsifiable concentrate.
Use: Selective, postemergent.
Rate: 0.25 to 0.5 lb ai/A in 50 to 100 gal of water. Use on cool season turf species only.

Remarks: Especially useful for creeping woodsorrel *(Oxalis)* control. Apply on vigorously growing broadleaf weeds, preferably in the spring or fall. May be retreated 4 weeks following the first application for hard-to-kill weeds. To broaden weed spectrum and control dandelion, use a tank mix of amine or low volatile ester of 2,4-D with triclopyr. Do not apply around trees or shrubs, since injury may result. Do not follow application with an irrigation within 4 hours.

2,4-D* low-volatile esters (Chacon, Broad-Leaf Weed Killer, Esteron 99 Concentrate, Weedone LV-4 [others])

Formulation: Emulsifiable.
Use: Selective, foliage-applied, translocated.
Rate: 2 lb ae in 100 gal water/A or 4 tsp formulation per 1 gal water for spot treatment.
Remarks: To control common yarrow, speedwells, mallows, mature knotweed.

2,4-D* oil-soluble amines (Dacamine, Emulsamine E-3)

Formulation: Emulsifiable.
Use: Selective, foliage-applied, translocated.
Rate: 1 lb ae in 100 gal water/A or 2 tsp of 2 lb/gal formulation/gal water for spot treatment.
Remarks: To control dandelion, plantain, young knotweed (2- to 4-leaf stage).
Do not exceed 1 lb ae/A on bentgrass. Some injury may result. Apply only on established turfgrass. Use in dichondra turf only as a nonselective spot treatment. Apply only on calm days to avoid drift.

2,4-D* water-soluble amines (Formula 40, Weedar 64, 2,4-D Amine No. 4, others)

Use: Selective, foliage-applied, translocated.
Rate: 1 lb ae+1 qt surfactant in 100 gal water/A or 2 tsp formulation+2 tsp surfactant to 1 gal water for spot treatment.
Remarks: To control dandelion, plantain, young pigweed.
NOTE: On bentgrasses use water-soluble amine only and do not exceed 3/4 lb ae/A.
Rate: 2 lb ae+1 qt surfactant in 100 gal water/A or 4 tsp formulation+2 tsp. surfactant to 1 gal water for spot treatment.
Remarks: To control young knotweed (2- to 4-leaf stage), field bindweed, wild lettuce, filaree.

*Restricted herbicide; permit required from County Agricultural Commissioner.

DISEASES

RECOMMENDED FUNGICIDES

Disease	Fungicidal control (Use rates and frequencies recommended by the manufacturer)	Disease	Fungicidal control (Use rates and frequencies recommended by the manufacturer)
Anthracnose	chlorothalonil mancozeb triadimefon thiophanate-methyl	Fusarium patch	benomyl fenarimol iprodione mancozeb thiophanate-methyl triadimefon vinclozolin
Dollar spot	anilazine benomyl chlorothalonil fenarimol iprodione mancozeb thiophanate thiophanate-methyl thiram triadimefon vinclozolin	Helmintho-sporium leaf spot	anilazine captan chlorothalonil
		Leaf blotch	iprodione mancozeb
		Melting out	maneb thiram
Fairy ring	Complete soil sterilization. Methyl bromide* Soil-wetting agents may be helpful.	Powdery mildew	triadimefon fenarimol†
		Pythium blight or grease spot	chloroneb etridiazole mancozeb metalaxyl propamocarb
Fusarium blight complex	Complete control with fungicides has not been attained in California. benomyl fenarimol iprodione mancozeb thiophanate thiophanate-methyl triadimefon Water fungicides in after application.	Red thread	chlorothalonil iprodione mancozeb triadimefon fenarimol vinclozolin

Continued on next page

RECOMMENDED FUNGICIDES—*Continued*

Disease	Fungicidal control (Use rates and frequencies recommended by the manufacturer)	Disease	Fungicidal control (Use rates and frequencies recommended by the manufacturer)
Rhizoctonia blight (brown patch)	anilazine benomyl captan chlorothalonil fenarimol iprodione mancozeb PCNB thiophanate thiophanate-methyl thiram triadimefon	**Smut, loose** **Smut, stripe**	Treat seed with captan or thiram. Fungicides used for stripe smut might be effective. benomyl fenarimol thiophanate thiophanate-methyl triadimefon Treat seed with thiram or captan.
Rust	Triadimefon is most effective. oxycarboxin Also helpful: anilazine chlorothalonil mancozeb	**Southern blight** **Spring dead spot**	PCNB (Water into the turf.) triadimefon fenarimol‡ Benomyl applied experimentally in the fall provides some control.
Seed rot and damping off	Treat seed with thiram or captan. Fumigate soil before planting with dazomet, metham sodium or methyl bromide.*	**Take-all patch**	fenarimol§ triadimefon§

*Permit for possession or use required from County Agricultural Commissioner.
†Powdery mildew is not listed on product label, but there are reports of effectiveness.
‡Registration on spring dead spot is pending in California.
§Take-all patch is not listed on this label, but there are reports of effectiveness.

COMMON, CHEMICAL, AND TRADE NAMES OF TURF FUNGICIDES

The following is a list of turf fungicides and the trade names under which they may be purchased at nurseries or supply houses. The chemical names are given in lowercase letters; the trade names, many of which are copyrighted, have their initial letters capitalized. Mercury and cadmium compounds are not included.

Some products not mentioned are known to give excellent results in University tests, but are not yet registered for commercial use, and therefore cannot be recommended.

anilazine = 4,6-dichloro-N-(2-chlorophenyl)-1,3,5-triazin-2-amine: Best Turf Fungicide, Dyrene.

benomyl = methyl 1-(butylcarbamoyl)-2-benzimidazolecarbamate: Tersan 1991, Scotts DSB Fungicide.

captan = N-trichloromethylthio-4-cyclohexene-1,2-dicarboximide: Orthocide, Captan.

chloroneb = 1,4-dichloro-2,5-dimethyoxybenzene: Terraneb, Scotts Fungicide II.

chlorothalonil = tetrachloroisophthalonitrile: Daconil 2787, Best Turf Disease Control, Turf Care, Scotts 101BS Fungicide, Ortho Liquid Lawn Disease Control.

dicloran = 2,6-dichloro-4-nitroanaline: Botran.

etridiazole = 5-ethoxy-3-trichloromethyl-1,2,4-thiadiazole: Koban, Terrazole.

fenarimol = -(2-chlorophenyl)- -(4-chlorophenyl)-5-pyrimidinemethanol: Rubigan.

folpet = N-(trichloromethylthio)phthalimide: Folpan, Folpet, Phaltan.

iprodione = 3-(3,5-dichlorophenyl)-N-(1-methylethyl)-2,4-dioxo-1-imidazolidenecarboximide: Chipco 26019, Scotts Fungicide VI.

mancozeb = coordination product of zinc ion and manganous ethylenebisdithiocarbamate: Fore, Best Multipurpose Disease Control.

maneb = manganese ethylenebisdithiocarbamate: Tersan LSR, Manzate 200.

metalaxyl = N-(2,6-dimethylphenyl)-N-(methoxyacetyl) alanine methyl ester: Subdue.

oxycarboxin = 2,3-dihydro-5-carboxanilido-6-methyl-1,4-oxathiin-4,4-dioxide: Plantvax.

PCNB = pentachloronitrobenzene: Funticlor, Terraclor, Scotts FF II, Turfcide.

propamocarb=propyl [3-(dimethylamino) propyl] carbamate monohydrochloride: Banol.

thiophanate=diethyl,4,4-*o*-phenylenebis(3-thioallophanate): Cleary's-3336, Proturf Systemic Fungicide.

thiophanate-methyl=dimethyl [(1,2-phenylene) bis (iminocarbonothioyl)] bis (carbamate): Fungo 50, Ropsin M, Scotts Systemic Fungicide, Scotts DSB Fungicide.

thiram=tetramethylthiuram disulfide: D&P, Spotrete, Turf-Tox, Thiuran 75.

triadimefon=1-(4-chlorophenoxy)-3,3-dimethyl-1-(1*H*-1,2,4-triazol-1-yl)-2-butanone: Bayleton, Scotts Fungicide 7.

vinclozolin=3-(3,5-dichlorophenyl)-5-ethenyl-5-methyl-2,4-oxazolidinedione: Vorlan.

Combinations

dazomet=3,5-dimethyl tetrahydro-1,3,5-2 thiadiazine-2-thione: Basamid.

methyl bromide*=methyl bromide: Bed Fume, Bromex, Brom-O-Gas, MBC Fumigant, Pestmaster Soil Fumigant, Tribrome, Weedfume.

MIT=methyl isothiocyanate: Vorlex (Vorlex is 20% MIT and 80% chlorinated hydrocarbon).

SMDC=sodium methyldithiocarbamate: Vapam, Soil-Prep.

---

*Permit required from County Agricultural Commissioner for possession or use.

INSECTS, MITES, AND MOLLUSCS

Insect and related pest damage to lawns often resembles the symptoms of diseases, nematodes, poor soil conditions, or other factors. Therefore, before you apply insecticides, it is usually best to make sure that insects are causing the damage.

HOW TO LOOK FOR INSECTS

Use the pyrethrum test. Mix 1 tablespoon of a commercial pyrethrum preparation (containing 1% to 2% pyrethrins) in 1 gallon of water and apply to 1 square yard of lawn. This brings vegetable weevil larvae, lawn moth larvae (sod webworms), cutworms, and other caterpillars to the surface within 10 minutes. It also brings up earwigs, but does not indicate whether white grubs or billbugs are present. Finding a few caterpillars is normal. Usually, no insecticide is needed unless there are more than 5 cutworms or 15 lawn moth larvae per square yard.

Select an area containing some living grass, and examine the soil around the roots for the white, legless larvae (grubs) of billbugs or the C-shaped, legged larvae of June beetles (white grubs). When abundant, these insects can eat away the roots of the grass so the turf can be rolled back like a carpet. If you find more than one billbug grub or one white grub per square foot, apply an insecticide.

For spider mites, leafhoppers, and flea beetles, examine the leaves, stems, and crowns of the plants. For snails and slugs, look for the dry mucous trails left by these pests.

CONTROL MEASURES FOR LAWN INSECTS, MITES, AND MOLLUSCS

Pest	Pesticide	Remarks
Lawn moths (sod webworms) (grasses)	Dursban spray or granules	Mow lawn and water well before treatment. Apply when plants are dry. With sprays, do not water again until necessary.
	Sprays with *Bacillus thuringiensis*	
	Orthene Soluble Powder	
	Proxol/Dylox spray	
	carbaryl (Sevin)*	
Cutworms and armyworms (grasses and dichondra)	Dursban spray or granules	Mow lawn and water well before treatment. Apply when plants are dry. Follow sprays with an equal amount of water to carry insecticide into plant crowns.
	Orthene Soluble Powder	
	Proxol/Dylox spray	
	carbaryl (Sevin)*	
Vegetable weevil, leafhoppers (grasses) **flea beetle** (a pest of dichondra during warmer months) **Fungus gnats, march flies**	malathion spray	Mow lawn and water well before treatment. Apply when plants are dry. Do not water again until necessary. Repeat applications may be necessary for flea beetles.
	Dursban spray (vegetable weevil and flea beetle only)	
	Orthene Soluble Powder (leafhoppers only)	
White grubs	Dursban spray or granules	Water heavily after application to wash insecticide into plant root zone. Do not wash insecticide away by flooding. Repeat applications may be necessary.
	Mocap* granules	
	Proxol/Dylox, Turcam	

Continued on next page

*Permit from County Agricultural Commissioner required for possession or use.

CONTROL MEASURES FOR LAWN INSECTS, MITES, AND MOLLUSCS—*Continued*

Pest	Pesticide	Remarks
Chinch bug	Mocap* granules	Water heavily after application to wash insecticide into plant crown. Do not wash insecticide away by flooding. Repeat applications may be necessary.
Snails and slugs	bait or granules containing metaldehyde or metaldehyde plus Sevin* or Mesurol	Apply in late evening. Control with baits is improved by sprinkling area lightly with water before treatment to activate snails and slugs.

*Permit from County Agricultural Commissioner required for possession or use.

COMMON, TRADE, AND CHEMICAL NAMES OF TURF INSECTICIDES

acephate = *O,S*-dimethyl acetylphosphoramidothioate: Orthene.

Bacillus thuringiensis **var.** *kurstaki:* several.

bendiocarb = 2,2-dimethyl-1,3-benzodioxol-4-yl methylcarbamate: Turcam.

carbaryl = 1-napthyl *N*-methylcarbamate: Sevin.*

chlorpyrifos = *O,O*-diethyl *O*-(3,5,6-trichloro-2-pyridyl) phosphorothioate: Dursban.

ethoprop = *O*-ethyl *S,S*-dipropyl phosphorodithioate: Mocap.

malathion = diethyl mercaptosuccinate, *S*-ester with *O,O*-dimethyl phosphorodithioate: Malathion, Cythion.

methiocarb **metmercapturon** = 3,5-dimethyl-4-(methylthio)phenyl methylcarbamate: Mesurol.

metaldehyde = metacetaldehyde: Metaldehyde.

trichlorfon = dimethyl (2,2,2-trichloro-1-hydroxyethyl)phosphonate: Dylox, Proxol.

NEMATODES

Make sure to investigate local restrictions on the handling and application of nematode-control chemicals. Read container labels and follow instructions carefully. Some of the materials listed may not be used without a permit from the County Agricultural Commissioner. Some are not available to the homeowner and must be applied by a licensed Pest Control Operator.

REGISTERED PREPLANT AND POSTPLANT CHEMICALS FOR NEMATODE CONTROL ON TURF

Chemical	Pests controlled*	Tarping required?	Application method
Preplant materials			
Methyl bromide†	1, 2, 3, 4	yes	inject from 1- to 1½-lb compressed gas canisters every 100 sq ft under polyethylene tarps
Chloropicrin†	1, 2, 3, 4	no	inject; preferably, cover with polyethylene tarps
SMDC†	1, 2, 3, 4	no	sprinkle on and water in, or apply through sprinklers
Dichloropropene	1	no	inject
Postplant materials‡			
Fensulfothion†	1	no	granules
Phenamiphos (fenamiphos)†	1	no	granules
Ethoprop†	1	no	granules

*1 = nematodes, 2 = fungi, 3 = insects, 4 = weeds.

†Restricted material; permit required from County Agricultural Commissioner.

‡None of the materials listed for postplant application is labeled for use on golf greens at this time.

COMMON, CHEMICAL, AND TRADE NAMES OF TURF NEMATICIDES

chloropicrin = trichloronitromethane.

SMDC = sodium N-methyldithiocarbamate: Vapam.

1,3-D = 1,3-dichloropropene: Telone.

methyl bromide = bromomethane.

phenamiphos (fenamiphos) = ethyl 3-methyl-4-(methylthio) phenyl (1-methylethyl) phosphoramidate: Nemacur.

ethoprop = O-ethyl S,S-dipropyl phosphorodithioate: Mocap.

RODENTS AND RELATED VERTEBRATE PESTS

Rodents and related vertebrate pests of turfgrass can cause severe damage and management problems. Early detection of these pests is important, since their presence often leads to damage. If uncontrolled, their populations will likely increase to intolerable levels.

Before control measures are taken, the animal believed to be the cause of the problem must be correctly identified. The resolution of the problem will depend on this. Use *Turfgrass Pests* (Publication 4053) to identify the pest, its sign (indicators of activity), and the type of damage done. Nonlethal and nonchemical control methods are effective in many instances. Long-term management of the turf and surrounding areas is often the best approach for prevention of rodent and related vertebrate pest damage.

WARNING

> The descriptions of toxicants and poison baits given here are intended as brief guides to their properties and uses. ALL ARE HAZARDOUS to some degree and should be handled carefully and kept in locked storage when not in use. In all cases, follow label instructions carefully. Some toxicants discussed are restricted-use materials. Permits from the County Agricultural Commissioner's Office are required for purchase or use of such materials. In some cases, when limited quantities are used, no permit is required. Check with the Agricultural Commissioner's Office to be sure.

GENERAL INFORMATION ON TOXIC BAITS

Many toxicants for turfgrass rodents and similar pests are formulated as baits on whole grains or pelleted cereals. To provide adequate control, a bait must be consumed in sufficient quantities. Care should be taken to use good-quality, fresh bait appropriate to the target pest. When using poison baits, take care to insure the safety of children, pets, and nontarget animals. Follow product label instructions carefully.

Multiple-feeding baits. Anticoagulant baits are both cumulative and slow-acting and thus must be consumed over a period of 5 or more days in order to be effective. They are probably the safest rodent baits for use around homes and other inhabited areas, although they must still be used with care. Many types and brands are available. Whole-grain baits are commonly used, but pelleted baits can also work well. Moisure-resistant paraffin block baits are useful along drainage ditches and in areas where moisture may cause other baits to spoil.

Because the pest must feed on most anticoagulant baits for 5 or more days, the bait must remain available until the population is controlled. As with trapping, bait placement is very important—the animal has to find and eat the bait. If you broadcast the bait, you will probably have to apply it every other day for three or four treatments.

Single-feeding baits. Strychnine and zinc phosphide are toxicants that kill after only one feeding. No retreatment with these materials is recommended (except strychnine for pocket gophers), since the animal often becomes bait shy after consuming a sublethal dose. All rodenticides (i.e., poison baits) should be considered hazardous to children, pets, and nontarget animals, and this is especially true of strychnine and zinc phosphide. Follow product label instructions carefully to avoid potential hazards.

FUMIGANTS

Chemicals in the fumigant group are used to control burrowing rodents (primarily ground squirrels). Effectiveness depends on the production and retention of a toxic gas concentration within the burrow system. These factors are affected by soil density, moisture, temperature, soil cracks through which gases escape, and burrow capacity. Application time, labor, and equipment usually limit the fumigation method to smaller infestations, or to follow-up for other methods. Some fumigants present the possibility of accidental fires, so caution is needed. Because all fumigants are toxic to people and animals, never use them beneath buildings.

CONTROL MEASURES FOR RODENTS AND RELATED PESTS

Pest	Rodenticide	Remarks
Ground squirrel *(Spermophilus beecheyii)*	anticoagulant (bait)	Most effective in early summer or fall, but can be used whenever squirrels eat the bait. Use in bait boxes or stations to prevent access to bait by children, pets, and nontarget wildlife, or broadcast by hand or by mechanical means according to label instructions. Squirrels may first enter bait stations as many as 4 to 10 days after application. Uninterrupted access to bait for 5 to 10 days is required for good control. When broadcasting bait, repeat applications as specified on the label. Population reduction may not occur for 2 to 4 weeks.

Continued on next page

CONTROL MEASURES FOR RODENTS AND RELATED PESTS—*Continued*

Pest	Rodenticide	Remarks
	strychnine (bait)	Restricted-Use Material. Most effective on squirrels north of Sacramento and San Francisco in early summer and fall, and in other areas during the fall when the squirrels pouch grain or seeds. Scatter bait near burrow openings according to label directions. Use only once per season.
	zinc phosphide (bait)	Restricted-Use Material. Most effective in early summer and fall when squirrels are active and taking grain baits. Scatter near burrow openings according to label directions. Use only once per season.
	aluminum phosphide (fumigant)	Restricted-Use Material. Use when squirrels are active aboveground. Most effective when soil is moist—during the spring or after irrigation. Place tablets in burrow opening, stuff with crumpled newspaper, and cover with soil. Treat all burrow openings. Retreat any newly opened burrows in 3 to 4 days.
	gas cartridge (fumigant)	Use when squirrels are active aboveground. Most effective when soil is moist, such as during the spring or after irrigation. Place cartridge in active burrow, ignite, push it as deep as possible with a stick, and cover the opening. Treat all burrows. Retreat newly opened burrows in 3 to 4 days.
Jackrabbit *(Lepus californicas)*	anticoagulant (bait)	Place bait in a bait station in an area frequented by rabbits (near trails or resting and feeding areas). Bait must be fed on for 5 days or more to be effective. Population reduction may not occur for 2 to 4 weeks.
	strychnine (bait)	Restricted-Use Material. Establish baiting sites about 100 yards from the turf area being damaged. Place prebait (unpoisoned bait) at 10- to 15-yard intervals, preferably along the rabbit trail. Prebait for 3 to 5 days to condition rabbits to feed at the baiting site. Once adequate feeding occurs, remove prebait and place bait at the same site. After 1 to 2 days of bait exposure, remove all uneaten bait. Do not treat more than once per season.
Meadow vole *(Microtus* spp.*)*	anticoagulant (bait)	Place in vole holes or runways or broadcast by hand or by mechanical means according to label. Repeat treatment every other day for 3 to 4 applications or as specified on label.

Continued on next page

CONTROL MEASURES FOR RODENTS AND RELATED PESTS—*Continued*

Pest	Rodenticide	Remarks
	zinc phosphide (bait)	Restricted-Use Material. Place in vole holes or runways or broadcast by hand or by mechanical means according to label. To avoid bait shyness, apply only once in a 4- to 6-month period.
	aluminum phosphide* (fumigant)	Restricted-Use Material. Use when soil moisture is high. Place tablets in all burrow openings and cover. Treat all active holes. To find active burrows, cover all openings with soil 7 to 10 days before treatment. Retreat newly open burrows in 3 to 4 days. Repeat applications may be necessary. Generally not used unless other materials have been ineffective.
Mole *(Scapanus* spp.*)*	strychnine* (bait)	Restricted-Use Material. Baiting for moles is considered relatively ineffective. Place bait underground into deep mole tunnels. Check for activity 3 to 5 days after treatment by knocking down mounds. Repeated treatments are often necessary.
	gas cartridge* (fumigant)	Place in the deeper underground burrow system, ignite, and plug entry hole. Check for new activity the following day by knocking down mounds and subsurface tunnels. Treat again if necessary.
	aluminum phosphide* (fumigant)	Restricted-Use Material. Place tablets in deep underground tunnels and plug the opening. Several placements per burrow system are necessary. Check for activity in 2 to 3 days by knocking down mounds and subsurface tunnels. Treat again if necessary.
Pocket gopher *(Thomomys* spp.*)*	strychnine (bait)	Restricted-Use Material. Place bait underground in gopher burrow using probe or mechanical bait applicator. After 3 to 5 days, knock down all mounds and recheck for new activity. Treat again as necessary.
	anticoagulant (bait)	Place bait underground in burrow system as above. Follow label recommendations on baiting quantity, placement, and number of repeat applications. Gopher activity will likely continue for 1 to 3 weeks after baiting because of the slow action of these materials.
	zinc phosphide* (bait)	Restricted-Use Material. Place bait underground in burrow system as above. After 3 to 5 days, knock down mounds to check for new activity.

Continued on next page

*Little or no information has been published on this material's effectiveness in California against this pest.

CONTROL MEASURES FOR RODENTS AND RELATED PESTS—*Continued*

Pest	Rodenticide	Remarks
	aluminum phosphide* (fumigant)	Restricted-Use Material. Use when soil moisture is high. Place tablets in gopher tunnel and seal opening. Several placements may be necessary for each burrow system. Check for activity by knocking down mounds 2 to 4 days after treatment. Repeat application may be necessary.
Skunk *(Spilogale putorius, Mephitis mephitis)* **and raccoon** *(Procyon lotor)*		No chemicals are registered as repellents or toxicants. Sometimes controlling turf grubs reduces the damage from skunks and raccoons since it reduces their food supply, causing them to move on. Fences can be effective.
Other: Chemicals registered but not recommended for use in turf		
	compound 1080	Highly restricted material for use by or under the direct supervision of government employees. Not generally applied in turf areas.
	methyl bromide	Restricted-Use Material. Place gas using special applicator in underground burrow. THIS MATERIAL IS PHYTOTOXIC, AND IS NOT RECOMMENDED FOR USE IN TURF. EXTENSIVE DAMAGE TO PLANTS MAY OCCUR.

PLANT
PESTICIDE USE WARNING — READ THE LABEL

Pesticides are poisonous and must be used with caution. READ the label CAREFULLY BEFORE opening a container. Precautions and directions MUST be followed exactly. Special protective equipment as indicated must be used.

STORAGE: Keep all pesticides in original containers only. Store separately in a locked shed or area. Keep all pesticides out of the reach of children, unauthorized personnel, pets and livestock. DO NOT STORE with foods, feeds or fertilizers. Post warning signs on pesticide storage areas.

USE: The suggestions given in this publication are based upon best current information. Follow directions: measure accurately to avoid residues exceeding tolerances, use exact amounts as indicated on the label or lesser amounts given in this publication. Use a pesticide only on crops, plants or animals shown on the label.

CONTAINER DISPOSAL: Consult your County Agricultural Commissioner for correct procedures for rinsing and disposing of empty containers. Do not transport pesticides in vehicles with foods, feeds, clothing, or other materials, and never in a closed cab with the vehicle driver.

RESPONSIBILITY: The grower is legally responsible for proper use of pesticides including drift to other crops or properties, and for excessive residues. Pesticides should not be applied over streams, rivers, ponds, lakes, run-off irrigation or other aquatic areas except where specific use for that purpose is intended.

BENEFICIAL INSECTS: Many pesticides are highly toxic to honey bees and other beneficial insects. The farmer, the beekeeper and the pest control industry should cooperate closely to keep losses of beneficial species to a minimum.

PROCESSED CROPS: Some processors will not accept a crop treated with certain chemicals. If your crop is going to a processor, be sure to check with the processor before making a pesticide application.

POSTING TREATED FIELDS: When worker safety reentry intervals are established be sure to keep workers out and post the treated areas with signs when required indicating the safe reentry date.

PERMIT REQUIREMENTS: Many pesticides require a permit from the County Agricultural Commissioner before possession or use. When such compounds are recommended in this publication, they are marked with an asterisk (*).

PLANT INJURY: Certain chemicals may cause injury or give less than optimum pest control if used at the wrong stage of plant development; in certain soil types; when temperatures are too high or too low; the wrong formulation is used; and excessive rates or incompatible materials are used.

PERSONAL SAFETY: Follow label directions exactly. Avoid splashing, spilling, leaks, spray drift or clothing contamination. Do NOT eat, smoke, drink, or chew while using pesticides. Provide for emergency medical care in advance.

To simplify information, trade names of products have been used. No endorsement of named products is intended, nor is criticism implied of similar products which are not mentioned.

Good fertilization, watering, and aeration programs cannot be overemphasized. Healthy, vigorous turf can outgrow the effects of an infestation, while poor growing conditions will result in more severe damage and slower recovery. In addition, the healthier the turf is, the fewer insecticide treatments will be required—if any are needed at all. Water-stressed turf stands a greater chance of being damaged by chemicals applied for pest control.

Turf is especially vulnerable to insect attacks when it is becoming established. To minimize the damage, closely monitor newly planted turf. Amend the soil and keep it fertilized to enhance growth so the turf will be less susceptible to damage. Keep the turf weed-free. Insects may first be attracted to the weeds in turf, and then move onto the turf itself. By eliminating unwanted weeds, you can reduce some of your insect problems and have a better-looking turf area.

INSECTICIDE APPLICATION

Lawns that require treatment to control pests that feed on grass leaves, stems, and crowns generally should first be irrigated well. As soon as the plants dry, apply the insecticide, and then withhold further irrigation until necessary to prevent wilting. This will allow the insecticide to remain on the plants for the longest possible period, maximizing the effectiveness of the treatment. When lawns are to be treated for pests such as white grubs or the larvae of billbugs that feed belowground on the grass roots, follow the application with a heavy irrigation to carry the insecticide into the soil.

Do not apply insecticides to water-stressed dichondra, or when the temperature exceeds 90°F.

In general, sprays are preferable to other application methods for pests of turfgrass and dichondra. Granular formulations, however, are also suitable for control of white grubs, billbugs, chinch bugs, cutworms, skipper larvae, sod webworms, slugs, and snails.

You can apply the required amount of insecticide indicated on the label in any convenient amount of water, provided you use enough water to thoroughly wet the grass or dichondra down to the ground. Many applicators use about 25 gallons of spray per 1,000 square feet. Control of spider mites may require a greater volume. (See chapter 2, Measurements, Calculations, and Sprayer Preparation.)

Insect and Related Pests of Turfgrass and Dichondra

PEST AND DESCRIPTION	LIFE HISTORY AND HABITS
AUSTRALIAN SOD FLY, *Inopus rubriceps* (Macquart). Adult males are ¼ inch long, black, with yellowish legs. Females are ⅜ inch long, black, with reddish legs and a red head. Larvae are ½ inch long, tan, with a dark head and dark, stiff setae.	Eggs laid in the soil hatch after 2 weeks. Larvae feed upon the roots and may live up to 2 years. Pupation occurs in the soil and lasts about a month. Adults live for a week, mate, lay eggs and die. Major adult flight activity can be expected in October. In some years, however, a minor flight may occur during May.
BERMUDAGRASS MITE, *Eriophes cynodoniensis* Sayed. Extremely small (can barely be seen with a 10x hand lens), white, worm-like mite. Has two pairs of legs near head end. Females lay spherical, transparent eggs singly or in groups. Also called Eriophyid mite or stunt mite.	This pest is thought to overwinter beneath leaf sheaths in the crown of bermudagrass. Females begin laying eggs beneath leaf sheaths of new growth in the spring. Breeding continues during the warm part of the year, and several generations occur during the season; a generation can be completed in 5 to 10 days. The mites suck juices from stems and within leaf sheaths.
BERMUDAGRASS SCALE, *Odonaspis ruthae* Kot. Adults are about 1/16 inch long and covered with a whitish, clam-shaped, waxy shield.	Bermudagrass scale commonly infests bermudagrass in southern California. It occurs primarily on the stems, but may also be on leaves. It is most likely to occur in lawn areas shaded by trees or buildings and is favored by the development of a heavy thatch. Females lay eggs under their waxy shield. Crawlers hatch, disperse for short distances, then become sessile and start feeding.

Adult Australian sod fly (Inopus rubriceps [Macquart]).

Turf damage from bermudagrass mite (Eriophes cynodoniensis Sayed.).

SUSCEPTIBLE PLANTS	DAMAGE	DETECTION
All grasses.	Larvae feed by extracting plant sap from the roots. The lawn becomes unthrifty, declines slowly, and is replaced by broadleaved weeds over a long period of time. Adults emerging in the fall are a nuisance in lawns when numerous. This insect has mainly been a pest of turf in the San Francisco Bay Area.	Inspect the root zone. Look for the tan, elongated larvae with dark heads. Adults can be seen on the grass from September through November. In occasional years, adults may also be seen in May.
Common and hybrid bermudagrasses.	Shortening of stem internodes, resulting in stunted, tufted, or rosetted plant appearance. With heavy infestations, the grass turns brown and dies. When infestations are allowed to persist, the grass may thin out, allowing the growth of weeds.	Look for plants with stunted, rosetted or tufted appearance. Pull leaf sheaths away from stems. Examine the inside of leaf sheaths and exposed stems with 10x or 20x hand lens or dissecting microscope. Look for mites and spherical, transparent eggs.
Bermudagrass.	Bermudagrass scales suck juices from the plants. Heavily infested grass takes on a brown, dry appearance, and new growth is retarded.	Close examination of infested grass will reveal a whitish material on the stems and crown that gives it a "moldy" appearance. Even closer examination with a hand lens or microscope will reveal the clam-shaped shields of the scale insects.

Continued on next page.

Bermudagrass scale (Odonaspis ruthae *Kot.*).

Insect and Related Pests of Turfgrass and Dichondra—Continued

PEST AND DESCRIPTION	LIFE HISTORY AND HABITS
BILLBUGS, *Sphenophorus phoeniciensis* Chitt. and *parvulus* Gyllenhal, *S. venatus vestitus* Chitt. The larvae of billbugs are white, legless grubs from 1/3 to 3/8 inch long when full grown. The adults are small black or brown weevils, or "snout" beetles.	No males are known for these species. Adult females can be found throughout the year, and in many areas larvae can also be found. Eggs are laid singly in adult feeding holes in grass stems. Larvae feed in the stem, crown, and roots, and then pupate in the soil. Adults emerge to repeat the cycle.
CLOVER MITE, *Bryobia praetiosa* Koch. Adults are about 1/30 inch in length, with long front legs. Bodies are somewhat depressed and reddish brown to greenish in color. The legs are amber to orange.	Eggs are laid on clover and most grasses. Feeding and reproduction occur in the cool periods of spring and fall. As temperatures rise, the mites become less active. Summer is spent in the egg stage. Several generations may occur during spring, fall, and winter.
CUTWORMS AND ARMYWORMS. *Agrotis ipsilon* (Hufn.), *Peridroma saucia* (Hubner), *Pseudaletia unipuncta* (Haw.), *Feltia subterranea* (Fab.). Thick-bodied caterpillars from 1 to 2 inches long when full grown. Usually dull colored, green, gray, brown or black; often with spots or longitudinal stripes. Adults are night-flying moths, dull or somber colored and with a wing span of 1 1/4 to 1 1/2 inches.	Moths fly at night and lay eggs on leaves of grasses, dichondra, or nearby plants. Larvae feed at night and hide in holes, under debris, or in mat of organic matter at the surface of the ground during the day. Pupation occurs in the soil. Breeding continues throughout the warm months of the year and there may be several generations per season with overlapping broods.

Billbug damage to lawn. Detail: the hunting billbug (Sphenophorus venatus vestitus).

SUSCEPTIBLE PLANTS	DAMAGE	DETECTION
All grasses. Old lawns in warm inland areas are most susceptible.	Young billing larvae or grubs feed inside the stems and crown. Older larvae feed beneath the ground on the roots of grasses. Aboveground symptoms are brown and dying grass in spots, or sometimes in large areas.	Whitish sawdustlike frass can be found on the ground as evidence of feeding by the larvae. Examine the soil around the grass roots. Dig in the edges of brown areas near green, healthy grass. If, on the average, more than 1 grub is found per square foot, the lawn should be treated.
Clovers. Most grasses.	Sucking out the plant juices during feeding results in a silvery appearance to the turf. Damage is most serious in areas close to buildings or planter boxes. Mites may invade houses looking for shelters. No reports of mites biting humans or pets are recorded. When crushed, the mites leave a red stain.	Close examination of the leaves with a hand lens or a microscope.
All grasses. Dichondra.	Cutworms feed on the leaves and crown and may cut off plants near the soil surface. Only the larvae are injurious.	Pyrethrum test. Treat when, on the average, 5 or more cutworms are found per square yard.

Continued on next page.

Cutworm (Agrotis ipsilon). Above: *larva.* Below: *adult.*

Insect and Related Pests of Turfgrass and Dichondra—Continued

PEST AND DESCRIPTION	LIFE HISTORY AND HABITS

FLEA BEETLE. *Chaetocnema repens* McCrea. Adult flea beetles are black and very small, about 1 millimeter ($1/25$ inch) long with extremely enlarged femora on the hind legs. The larvae live in the soil and are not usually seen.

Adults appear to be active from May through October in warm areas. One generation is completed in about a month. Eggs are laid on leaves and stems. Larvae develop on the roots of plants and pupate in the soil. Adult beetles jump readily when disturbed, hence the name "flea beetle."

FRIT FLY, *Oscinella frit* (L.). Adult frit flies are slightly more than $1/16$ inch long, shining black with small yellow markings on the legs. The eggs are pure white, $1/32$ inch long, with a finely ridged surface. Mature larvae are $1/8$ inch long, yellow, with black, curved mouth hooks. Pupae are yellow at first, then turn dark brown and are slightly less than $1/8$ inch long.

The winter is passed in the larval stage in the stems of grasses. Pupation takes place in the spring, and the first adults emerge about March. Eggs are laid on the leaves and leaf sheaths of grasses. Several larvae may occur in one plant. There are at least three broods, the activity of the last extending into October in warmer areas.

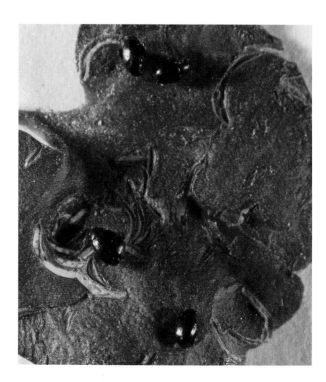

Flea beetles (Chaetocnema repens McCrea) and damage done to dichondra.

SUSCEPTIBLE PLANTS	DAMAGE	DETECTION
Dichondra.	Adult flea beetles feed on upper leaf surfaces, causing crescent-shaped scars. When enough of the leaf is eaten away, the leaf turns brown. Injury may be localized or spotty, and damaged areas are often assumed to have been caused by lack of water or fertilizer burn. Larvae feeding on the roots usually cause the most damage.	Close inspection of infested dichondra will reveal the small, shiny, black beetles sitting on the leaves. Moving the open palm of the hand slowly over the surface of the plants will cause the beetles to jump about and some will land on the back of the hand or arm and can thus be readily seen.
All grasses. Bent- and bluegrasses are most susceptible to injury.	The larvae tunnel in the stems near the surface of the soil, causing the upper portion of the plant to turn brown and die. Damage is most common on golf greens. Injury appears first on the collars of the greens and moves in toward the center. The high or upper sections are usually the first to show the symptoms. Greens with high organic matter content appear to be most susceptible.	Look for small, black adult flies hovering close to the grass from mid to late morning. Look for the larvae in the stems near the ground level. A hand lens or dissecting microscope are useful in finding the very small larvae.

Continued on next page.

Bermudagrass damaged by frit fly larvae (Oscinella frit [L.]). Detail: *adult frit flies.*

Insect and Related Pests of Turfgrass and Dichondra—Continued

PEST AND DESCRIPTION	LIFE HISTORY AND HABITS
GROUND PEARLS, *Margarodes meridionales* Morr. These scale insects are 1/16 to 1/8 inch long. Females are wingless, with a pearly white or yellow waxy coat. Eggs are pinkish white, laid in a sac which may be attached to the female's body. Reproduction is usually parthenogenetic and males are rare.	Females lay eggs that hatch in June or July. Crawlers are slender and active. They settle on a rootlet, insert their piercing mouthparts and begin feeding. They may be found at depths of 8 to 12 inches. It is believed that 2 years are required to complete a generation.
LAWN MOTHS (SOD WEBWORMS), *Crambus sperryellus* Klots and *Tehama bonifatella* (Hulst). Larvae are slender, grayish, black-spotted caterpillars about 3/4 inch long when full grown, rather sluggish in their activity except when disturbed. Moths are whitish or buff-colored with a wingspan slightly over 1 inch. *C. sperryellus* has a white or silver stripe along the margin of its forewing. Wings fold close to body when at rest, giving slender appearance.	Moths hide during day in grass and shrubbery, fly over grass at dusk, and drop their eggs on the leaves. Larvae hatch in a few days and begin to feed. They feed at night and hide during the day in tunnels constructed of bits of grass and debris and lined with silk (hence the name webworms). They pupate in the soil/thatch layer. Moths begin flying in April or May in warm areas and breeding continues through October. *C. sperryellus* is found in inland dry areas and *T. bonifatella* in coastal areas. Up to 4 generations per year may occur in southern California.

Ground pearls (Margarodes meridionales Morr.) feeding on bermudagrass roots.

SUSCEPTIBLE PLANTS	DAMAGE	DETECTION
Bermudagrasses in the internal desert areas. Other hosts reported include St. Augustinegrass, centipedegrass, and Zoysiagrass.	By feeding on plant juices in the root zone, ground pearls cause the turf to wilt and appear unthrifty in irregular patches. Damage appears more in sandy than clayey soils. Control is usually not needed if proper fertilization and management keep the turf in a vigorous condition.	Soil core samples must be taken at depths of 4 to 12 inches. Inspection of the soil and roots will reveal the cystlike insects.
All grasses. White clover. Bent- and bluegrasses are most susceptible to injury, especially new lawns. Dichondra is not attacked.	Larvae feed on grass blades, growing tips, and greener portions of crown, but not on the roots. Damaged areas appear as scattered, irregular, brown patches in turf.	Pyrethrum test. Treat when an average of 15 or more worms are found per square yard. Preventive treatments are suggested for newly planted lawns.

Continued on next page.

Lawn moth larvae (Tehama bonifatella [*Hulst.*]).

Lawn moth adult (Tehama bonifatella [*Hulst.*]).

59

Insect and Related Pests of Turfgrass and Dichondra—Continued

PEST AND DESCRIPTION	LIFE HISTORY AND HABITS
LEAF BUGS, FLEAHOPPERS, etc. (Family Miridae). A common species is the white-marked fleahopper, *Spanogonicus albofasciatus* (Reut.). Adults are about ⅛ inch long, blackish or grayish with white markings on the wings, which are folded flat over the back. Their long antennae and white markings distinguish them from flea beetles.	Little is known about the life cycle on turfgrasses, but several generations per season are thought to occur in warm areas. Fleahoppers jump readily or fly short distances when disturbed.
LEAFHOPPERS, *Draeculacephala minerva* Ball, *Deltacephalus sonorus*, Ball, and others. Small, active insects ⅛ to ¼ inch long. Adults whitish green, yellow, or brownish gray. May be speckled or mottled. Adults fly or jump short distances when disturbed. Nymphs are various colors, as are adults, but lack wings. Nymphs have the characteristic habit of moving sideways or backward when disturbed.	Adult females insert eggs into tissues of the host plant. Eggs hatch in a few days in warm areas. Several generations may occur in a season, with 12 to 30 days required for each generation.
LUCERNE MOTH, *Nomophila noctuella* (D. and S.). Larvae are slender, spotted caterpillars similar in appearance to lawn moth larvae, but larger. Full-grown caterpillars are about 1 inch long with a tendency to wriggle actively when disturbed. Adult moths with wing span of an inch or slightly more. Hind wings are gray, forewings mottled gray-brown, with two pairs of indistinct dark spots.	Moths fly at night and females lay eggs on clover and other legumes as well as turfgrasses and dichondra. Activity is greatest during the warm months.
OXALIS MITE *Tetranychina harti* (Ewing). Adults are bright red and approximately the size of the two-spotted spider mite, but with conspicuously longer legs.	The mites live and feed on oxalis. No grasses are known to be affected.

Lucerne moth larvae (*Nomophila noctuella* [D. and S.]).

SUSCEPTIBLE PLANTS	DAMAGE	DETECTION
All grasses. Dichondra.	Both adults and nymphs suck juices from the leaves and stems of grasses or dichondra. Heavy infestations retard growth and cause the grass to die in spots.	Look for fleahoppers by running the open palm of the hand slowly over the grass. If these insects are present, they can be seen hopping about. Some may hop onto the back of the hand or onto sidewalks or curbs where they are easy to see.
All grasses.	Both adults and nymphs suck juices from the leaves and stems of grasses. Symptoms are the bleaching or dying out of the grass.	Examine areas in the lawn that look bleached or dry, and the surrounding green areas. Look closely for the small, wedge-shaped, jumping or flying adults as you walk. Also look closely on the leaves and stems for the nymphs, which move sideways or backward when disturbed.
All grasses. Clovers. Dichondra.	Larvae feed on both leaves and stems, which weakens plants and reduces their photosynthetic ability. Damage usually appears in late summer.	Pyrethrum test. Control measures are seldom needed.
Oxalis.	This mite sucks out the plant's juices and turns the oxalis white.	Close examination of the leaves with a hand lens or a microscope.

Continued on next page.

Insect and Related Pests of Turfgrass and Dichondra—Continued

PEST AND DESCRIPTION	LIFE HISTORY AND HABITS
RHODESGRASS MEALYBUG. *Antonina graminis* (Mask.). This pest was known as the Rhodesgrass scale. Adults are about 1/16 inch in diameter, globular, dark purplish-brown in color, and with a white cottony covering.	No males are known for this species. Females give birth to live young. Immatures crawl upwards, settle under leaf sheaths, and start feeding. Several generations are thought to occur per season, each requiring about 60 to 70 days for completion.
SKIPPERS. *Hylephila phyleus* (Drury), (Fiery skipper). Larvae about 1 inch long, and brownish yellow in color. First 2 body segments behind the head are smaller than the rest, giving the appearance of a "neck." Adults are small butterflies with wing span of a little more than 1 inch. Males have orange-yellow wings spotted with black; wings of the females are dark brown, with orange-yellow spots.	Female butterflies lay eggs singly on turfgrass leaves, as well as on other ornamental plants, during the warm months. Larvae feed in the thatch layer and pupate near the soil surface. Adults feed on nectar from various flowers. Lantana is a favorite. There are 3 to 5 generations per year.
SNAILS AND SLUGS. Shiny molluscs with protective shells (snails) or without (slugs).	Slugs and snails move on a slime trail and are more active under moist conditions. They hide under vegetation or in dark places during the day, coming out to feed at night.

Skipper larva (Hylephila phyleus [*Drury*]).

SUSCEPTIBLE PLANTS	DAMAGE	DETECTION
Rhodesgrass. Bermudagrasses. St. Augustine-grass. Tall fescue and centipedegrass.	Rhodesgrass mealybug sucks juices from the plant and reduces growth and vigor.	Rhodesgrass mealybug is detected in the same manner as bermudagrass scale—by close examination of the grass.
All grasses, especially bentgrasses. Dichondra is not usually attacked.	Larvae feed on leaves and crown.	Pyrethrum test. Treat when an average of 10 or more larvae are found per square yard.
Dichondra. Several ornamental and garden plants.	Damage is frequently more severe near flower beds, ground cover areas, shrubbery, or planter boxes where slugs and snails hide during the day. Prolonged feeding in these areas may leave only bare stems or may completely kill the dichondra planting.	Look for the slime trails, which appear shiny in the sunlight. Slugs and snails may also be seen in the evening or night with the aid of a flashlight, as they move onto lawn areas from daytime hiding places. Sprinkling water on suspected infested areas activates these pests and aids in detection.

Continued on next page.

Insect and Related Pests of Turfgrass and Dichondra—Continued

PEST AND DESCRIPTION	LIFE HISTORY AND HABITS
SOUTHERN CHINCH BUG, *Blissus insularis* Barber. Adults are about 1/5 inch long, black with nearly all-white wings, which are folded flat over the body. There are both long- and short-winged forms. The young are bright red but turn black as they approach adulthood.	There appear to be at least two generations a year, with all stages present in any month. The highest populations occur during the summer, and at this time development from egg to adult takes about 6 weeks.
SPIDER MITES, Tetranychidae. Adults are about 1/50 inch long; globular in shape; and reddish, yellowish, or greenish in color. The two-spotted mite has a pronounced dark spot on each side of the body.	Spider mites lay eggs on grasses, dichondra, oxalis, or other plants in lawns. Some species spin fine webbing that may entirely cover the grass in spots, especially in protected areas near curbs, walls, or planter boxes. Spider mites breed throughout the warm months of the year, producing several generations per season.
VEGETABLE WEEVIL, *Listroderes costirostris obliquus* (Klug.). Larvae are small, green, legless grubs about 3/8 inch long when fully grown. Adults are brownish or grayish weevils or "snout" beetles about 3/8 inch long. Wing covers are very rough or punctated with sparse, short setae or hairs. Each wing cover has a short, pointed protuberance on the top near the back end. The adults do not fly. Both larvae and adults are very slow and sluggish in movement.	Vegetable weevils are active only during the winter and spring months. No males are present and females produce young without mating. Both adults and larvae feed at night. Occasional adults can be seen during the day, but usually hide among plant foliage or in other dark places. Eggs are laid on plants or in soil during the fall. Larvae hide in the soil beneath the plants during the day.

Southern chinch bugs (Blissus insularis Barber). Left to right: late-instar nymphs and winged adult.

Two-spotted spider mites (Tetranychus urticae Koch.).

SUSCEPTIBLE PLANTS	DAMAGE	DETECTION
Although bermudagrass and zoysia may be infested, only St. Augustine grass is seriously damaged.	Yellowish to brownish patches in lawn, grass dries up and dies.	Close examination of the border between damaged and green areas will reveal adults and nymphs. Flotation method or pyrethrum test also may be useful.
All grasses. Dichondra. Clovers. Oxalis.	Mites feed by puncturing the leaves and sucking out the juices. First symptoms of this injury appear as a speckling or stippling of the leaves. This is followed by a yellowing, bronzing, or bleaching and drying of the leaves.	Close examination of the leaves for stippling, speckling, yellowing, bronzing, bleaching, drying, webbing, plus the presence of mites and mite eggs. A hand lens or dissecting microscope may be necessary to see the mites and eggs.
Dichondra.	Damage first appears as small holes in the leaves, but in heavy infestations the leaves may be skeletonized or completely removed leaving only the bare stems. Since the adults do not fly, infestation of new areas usually takes place very slowly and damage is usually localized or spotty.	Pyrethrum test. Adult beetles often crawl to nearby lights at night.

Continued on next page.

Vegetable weevils (Listroderes costirostris obliquus [*Klug.*]). Detail: *larvae.*

Insect and Related Pests of Turfgrass and Dichondra—Continued

PEST AND DESCRIPTION	LIFE HISTORY AND HABITS
WHITE GRUBS, *Cyclocephala* spp. Larvae or grubs about 1 to 1½ inches long when fully grown, C-shaped when at rest, with many folds or wrinkles in the front half of the body. Rear end of body often slightly larger in diameter than rest and may be bluish or blackish in color. A white grub has a brown head capsule and three pairs of conspicuous legs. (Grubs of the billbug and vegetable weevil, described elsewhere in this handbook, are legless.)	Common species of white grubs attacking turf in California have a 1-year life cycle. Adults emerge from the ground in late May and in June, and lay eggs in the grass root zone which hatch into C-shaped grubs. After 2 molts, mature 3rd instar grubs overwinter. They pupate in the spring, and adults emerge to repeat the cycle.
Adults of white grubs are commonly called May beetles, June beetles, or June bugs. They are a little over ½ inch long and yellowish brown with a dark brown head.	

White grub larvae (Cyclocephala *spp.*).

SUSCEPTIBLE PLANTS	DAMAGE	DETECTION
All grasses. Bermudagrass and ryegrass are most susceptible.	White grubs feed beneath the soil surface on the roots of grasses. Aboveground symptoms are browning and dying of the grass in localized spots or in large irregularly shaped areas. Where infestations are heavy, the grass roots may be entirely eaten away so the turf can be rolled back like a carpet. Damage is usually most severe in September and October when grubs are reaching maturity and growth of bermudagrass is slowing down. Little feeding appears to take place during the winter or early spring.	Examine the soil around the grass roots. Dig in brown areas near the edge of green, healthy areas of grass. If, on the average, more than one grub is found per square foot of area, the lawn should be treated.

Bermudagrass damage from white grubs (Cyclocephala *spp.*). *Detail: adults.*

Activity and Treatment Periods for Turf Pests

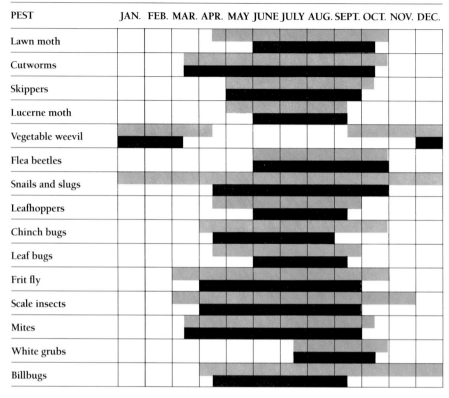

NOTE: The periods of pest activity indicated in the calendar are only approximate. Actual pest appearance will vary from area to area and even in the same areas from year to year, depending upon temperature, rainfall, and other factors, both physical and biological.

Treatment periods indicated are also only approximate. In most cases, insecticides should be applied only when pests are present in sufficient numbers to cause damage. An exception to this rule is the application of preventive treatment for control of lawn moths in newly planted lawns.

5

Nematode Diseases

John D. Radewald, extension nematologist,
university of california, riverside
Becky B. Westerdahl, extension nematologist,
university of california, davis

The parasites known to cause plant diseases are higher plants (mistletoe, dodder, broomrape) as well as insects, nematodes, fungi, bacteria, and viruses. A parasite is a living organism that obtains its food from another living organism; when it causes disease, it is pathogenic. Frequently, two or more pathogens act simultaneously on a plant in what is known as a "disease complex." However, simply to find any one of these potentially parasitic, disease-causing organisms within the root system of a given turf and assume guilt by association—that is, assuming (1) that the turf is diseased and (2) that the organism is a pathogen—is to make a serious error. Such associations only mean that turf disease may eventually occur, depending not only on the nematode, but also on the host and the environment. True pathogenicity can be proved only by reliable scientific methods, usually in the laboratory.

Some nematode species are injurious to turf in other parts of the United States, but none of these nematodes are known to occur on turf in California. Therefore, the turf grower in California has little information on which to base a decision regarding a nematode control program. However, investigations on this problem are underway.

WHAT IS A NEMATODE?

A nematode is a roundworm, sometimes called an eelworm or nema. Most plant-parasitic nematodes are microscopic and can only be identified by a trained technician using specialized equipment.

A nematode feeds by inserting its hollow spear (stylet) into the plant cells and sucking out the cellular contents. Some nematodes inject substances into the plant before they feed. The nematode's esophagus contains a median

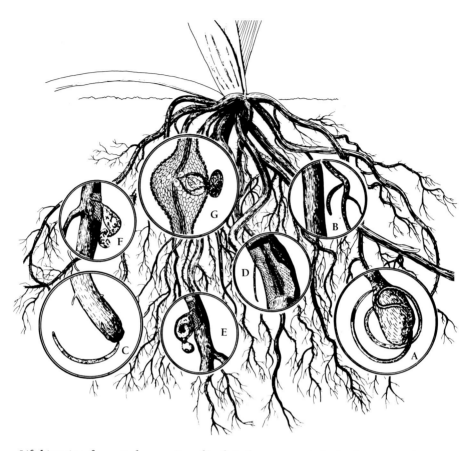

Life histories of nematode parasites of turf: A, *dagger nematode;* B, *ring nematode;* C, *stubby root nematode;* D, *lesion nematode;* E, *cyst nematode;* F, *reniform nematode;* G, *root-knot nematode.*

bulb to aid in the ingestion of plant cell contents and glands to produce digestive materials. The gut is a simple tube that terminates at the anus.

The female nematode lays eggs either singly or in egg masses, depending on the species. A single female of some nematodes may lay 500 eggs or more. A plant-parasitic nematode, being an obligate parasite, can complete its life cycle only when the roots or other portions of a susceptible host plant are present to feed upon.

Life histories, feeding habits, and methods of reproduction vary among nematodes. Some nematodes (ectoparasites) are migratory and feed strictly from the outside of the host plant. They deposit their eggs singly in the soil. Other migratory nematodes, such as the lesion nematode, feed with their bodies either inside or outside the root tissues, and deposit their eggs singly in the soil or within the root. Still other nematodes only migrate when young and become sedentary as adults. Their bodies may be outside the root or partially or completely embedded within the root. These nematodes lay egg masses in the soil. The eggs are surrounded by a gelatinous matrix. The egg mass of the root-knot nematode may be found within the root tissue with the adult female, or it may protrude from the root tissues into the soil. The female cyst nematode usually does not deposit her eggs, but retains them within her body. When the female dies, her body wall forms a leathery cyst that protects the eggs from adverse environmental conditions until they hatch.

When considering the application of a control chemical to established turf, you must determine whether the nematode and its eggs or young are located inside the root, outside the root, or in the soil. Nematodes and eggs inside the root tissues are protected from chemicals that are drenched or injected into the soil, unless the nematicide is systemic.

Nematode larvae molt four times. After the last molt, the nematode is an adult, and may vary in shape from the typical worm to a spherical female. The time required for a nematode to complete its life cycle from egg to adult varies. Under optimum conditions, most plant-parasitic nematodes complete this cycle in 3 to 6 weeks. The life cycle may extend to several months during periods of high or low temperatures or drought, or where hosts are poor.

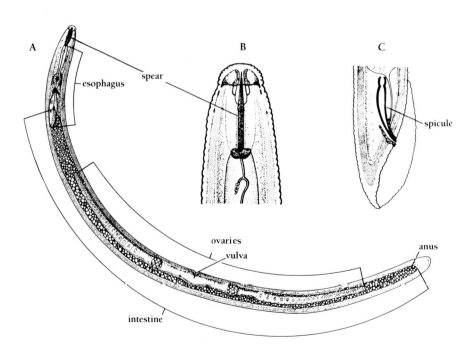

A. *Adult female nematode (actual length of nematode is 1/30 inch);* B. *Detail of head;* C. *Detail of male tail.*

PARTS OF TURF PLANTS ATTACKED

Nematodes are known to attack and damage most parts of the turf plant. Some nematodes destroy the seeds, while others attack the stems and leaves. The most frequently attacked portion of the turf plant is the root. There are no aboveground symptoms that exclusively indicate nematode damage to turf roots. Nematode-diseased plants may appear stunted, chlorotic (yellowed), and generally unthrifty. They also may wilt prematurely during the warmer periods of the day, because the root system that supplies the plant with water may be inadequate. The symptoms on the roots vary depending upon the nematode species. The most easily recognized nematode symptom on roots are the galls caused by the root-knot nematode, which are visible to the naked eye. The root-knot nematode is of great economic importance in California vegetable, field, and deciduous crops, and is commonly found on dichondra. In 1965, a new root-knot species, *Meloidogyne naasi,* was described. This nematode, found in two areas in California, is a proven pathogen of turfgrasses. Unlike most root-knot species, *M. naasi* seems to prefer grasses to other hosts.

The lesion nematode, *Pratylenchus* sp., is often found associated with turfgrasses, but is not a proven pathogen to turfgrasses in California. On other hosts, the symptoms are root lesions of various sizes and shapes.

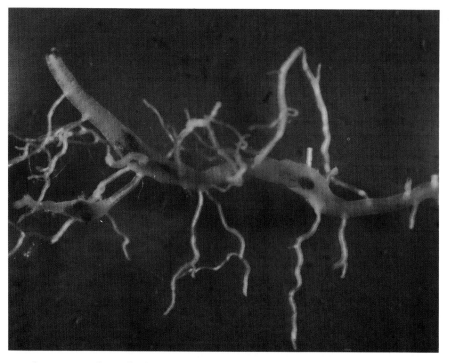

Root-knot nematode (Meloidogyne naasi) *on seaside bentgrass.*

Other nematodes, such as the stubby root nematode, attack only the growing root tips and stop their elongation. These root tips may appear swollen, discolored, or both, and growth of lateral roots is pronounced on some hosts. This nematode is commonly found around turf roots, but its importance on turf species in California is questionable. Other nematodes commonly found in association with turfgrasses in California include the ring nematode, *Criconemoides* sp.; the dagger nematode, *Xiphinema* sp.; and the pin nematode, *Paratylenchus* sp. Their detrimental effects to turf have not been studied in sufficient detail to determine their economic importance to California turf species.

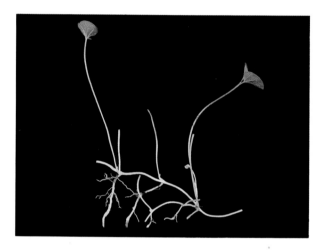

Root-knot nematode (Meloidogyne incognita) on dichondra.

DETECTING PLANT PARASITIC NEMATODES

The only way to detect a potential nematode disease of turf roots is by properly processing a soil and root sample. A sample consists of several soil cores, 1 to 2 inches in diameter and 6 to 8 inches in depth, taken in the area where the nematode problem is suspected. Include the turf roots with the soil. Place all soil cores from the suspect area in a single plastic bag and seal it to prevent drying. Take similar samples in adjacent apparently problem-free turf so you will be able to compare the nematodes found in the two areas. Label the sample bags carefully to maintain the identity of each individual sample. Keep the samples cool and transport them to a diagnostic laboratory as soon as possible, preferably the same day the samples are taken. Remember, *keep the samples cool and don't allow them to dry out.*

Your local farm advisor can give you a list of commercial laboratories in California that will test for and identify nematodes. Most laboratories provide a quantitative measure of the plant parasitic genera in each sample.

CONTROL

Plant-pathogenic nematodes on turfgrasses are controlled by the same methods used for their control on most other crops: quarantine and preplant or postplant measures.

Quarantine. Nematodes move very slowly in soil, possibly only a foot or two a year under their own power. Humans are the major disseminators of nematodes. In a turf nursery or landscaped area, nematodes spread easily with cultivation, watering, and the transfer of contaminated plants within a landscaped area. The first precaution, therefore, is to avoid introducing nematodes into uncontaminated areas. The turf grower may employ a self-imposed quarantine to avoid introducing nematode-infested soil or rootings into areas free of the pest. This self-imposed quarantine by the turf grower, landscape gardener, or homeowner is the simplest and most economical means of controlling potential nematode problems, provided the land is free of the pest.

Preplant control. When a turf nursery is to be established in barren soil or when the soil around a newly constructed home is infested with nematodes that are pathogenic to turf, a preplant nematode control program may be used. The compounds listed in Leaflet 2209 (*Guide to Turfgrass Pest Control*) will control nematodes. Some compounds will control other potential pests as well, including weeds, fungi, bacteria, and insects. When considering the large investment required to establish a turfgrass nursery or planting turf around a home, the most practical approach would be to control all potential pests possible before introducing the desired plant species, even though the initial cost of treatment may be high. The labor costs for weeding alone, as an example, in a newly established turfgrass nursery or home lawn can easily offset the initial cost of a broad-spectrum biocide such as those listed in Leaflet 2209. If nematodes are your only concern, a specific nematicide may be used for a reduced cost.

Postplant control. Some states make postplant recommendations for the control of various nematode diseases of turf. However, because research data in California are incomplete, no University recommendations can be made for postplant turf treatments for nematode control. If nematodes are detected in an established turf and the grower wishes to attempt a postplant control program, small areas may be treated with registered chemicals and compared with nontreated areas for plant response and actual nematode control. Your local farm advisor, University of California Cooperative Extension nematology specialist, or Experiment Station nematologist can suggest compounds for control measures and provide assistance in evaluating the experimental treatments.

6

Fungal Diseases

ARTHUR H. MCCAIN, EXTENSION PLANT PATHOLOGIST,
UNIVERSITY OF CALIFORNIA, BERKELEY
ROBERT M. ENDO, PROFESSOR OF PLANT PATHOLOGY,
UNIVERSITY OF CALIFORNIA, RIVERSIDE
HOWARD D. OHR, EXTENSION PLANT PATHOLOGIST,
UNIVERSITY OF CALIFORNIA, RIVERSIDE

Most of the serious turfgrass diseases in California result from infection by microscopic fungi. Fungi are low forms of plant life. Threadlike, devoid of chlorophyll, and incapable of manufacturing their own food, they live off dead or living plant or animal matter. Most fungi produce spores (seedlike bodies) and other structures resistant to adverse conditions. For example, sclerotia may be spread by wind, water, mechanical means such as mowers, or infected plant material such as grass clippings.

Spores, sclerotia, and vegetative threads (hyphae) require moisture and a favorable temperature to form new fungus threads, which grow over the surface of the plant or within it. These requirements mean that fungal diseases of grasses are most common during the rainy season and when irrigation moisture or dew remains on the leaves for long periods. Since fungi cannot develop without moisture, proper irrigation is the key to control. water early in the morning rather than late in the evening; deep, infrequent watering is better than shallow, frequent watering. Some diseases are favored by high temperatures, while others are most destructive during the cooler months.

Most turfgrass diseases are easier to prevent than to cure. To minimize the possibility of disease, plant the right kinds of grasses for your particular climatic zone. Weakened, ill-adapted grasses are susceptible to certain fungi and to stresses such as drought and hot, dry winds.

Recommended cultural practices such as mowing, fertilization, irrigation, aerification, and the like will help prevent diseases by maintaining a vigorously growing turf. A properly maintained turf generally sustains less severe damage from diseases and is able to recover more quickly than one that is poorly managed.

One cultural practice that is closely tied to disease prevention is fertilization—specifically, the amount of nitrogen applied. Too much nitrogen can result in soft, lush growth of the grass, a condition favorable to some diseases.

On the other hand, turf that has insufficient nitrogen will be susceptible to certain other diseases.

When you cannot obtain disease control by cultural methods, certain fungicides are recommended. Fungicides are usually most effective if applied before the disease becomes severe; later treatments will require higher rates and more frequent applications.

Every year, new and improved products become available. Therefore, we mention no specific fungicides here. Certain recommendations for chemical control of specific turfgrass diseases can be obtained from the University of California farm advisor in your area and from UC Division of Agriculture and Natural Resources Leaflet 2209 (*Guide to Turfgrass Pest Control*).

ANTHRACNOSE

Symptoms. Irregular patches of diseased turf 2 to 12 inches in diameter. Leaf blotches are brown, fading to light tan. Fungus forms minute black fruiting structures (acervuli) on dead grass blades. All turfgrasses are susceptible.

Cause. Anthracnose is caused by *Colletotrichum graminicola*. It is most severe under high temperatures (80° to 90°F), wet conditions, and low soil fertility.

Control. Apply adequate, balanced nutrients. Reduce irrigation frequency, consistent with vigorous growth. Several fungicides may be helpful when the disease is particularly severe.

DOLLAR SPOT

Symptoms. Small, circular areas of turf about 2 inches in diameter are affected. Spots may merge to form large, irregular areas. Leaves are water-soaked at first, later turn brown and, finally, straw-colored. Fine, white, cob-webby fungus threads may be seen in early morning on the infected grass.

Cause. *Sclerotinia homoeocarpa,* a fungus that survives in the soil by means of sclerotia, is the causal agent. The disease is common near or along the foggy coast, especially on bentgrass. Moderate temperatures (60° to 80°F), excess moisture, and excess mat and thatch favor dollar spot. Nitrogen-deficient turf develops more dollar spot than adequately fertilized turf.

Control. Keep thatch to a minimum. Water only when needed and to a depth of 4 to 6 inches. Apply adequate nitrogen. Fungicides are usually necessary to control this disease, especially on closely clipped grass such as golf greens. The fungicides are most effective if applied in the early spring and fall before disease develops.

Anthracnose damage to bentgrass (Agrostis spp.) and annual bluegrass (Poa annua).

Anthracnose damage to bentgrass (Agrostis spp.) and annual bluegrass (Poa annua) leaf tips.

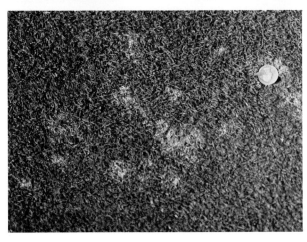

Dollar spot damage to bentgrass (Agrostis spp.).

FAIRY RING

Symptoms. A dark green band of turf develops in a circle or semicircle; mushrooms may or may not be present. Frequently, just behind the dark green band is an area of sparse, brown, dying grass, caused by lack of water penetration. Weed invasion is common. All turfgrasses can be affected by fairy ring, but the effects may be less on grass species with long stolons.

Cause. Several species of mushroom-forming fungi cause fairy ring. In northern and central California, the predominant fungus is *Marasmius oreades*; in southern California, *Lepiota* spp. Fairy ring develops most frequently in soil that is high in organic matter, such as undecomposed thatch.

Control. Apply adequate nitrogen. Remove thatch periodically. Aerate soil to improve water penetration and apply water heavily in holes every day for 3 to 10 days. It is possible to eliminate fairy ring fungus from soil by removing turf and fumigating the soil with volatile fumigants. However, this is a dangerous, expensive, and involved process and should be done *only* by a licensed specialist. Surfactants (wetting agents) may help water penetrate through the area affected by fairy ring.

FUSARIUM BLIGHT COMPLEX

Symptoms. The disease symptoms appear in hot weather during summer and early fall as small, circular areas of a few inches to a foot in diameter in which all or nearly all the plants within the circle have a root rot. The crown or basal area of the dead stems appears brown or black, and this tissue is hard and tough. The dead foliage becomes bleached in appearance. The disease principally attacks certain bluegrass cultivars, including *Poa annua*.

Cause. The fungus *Fusarium roseum f. cerealis* survives in the soil and turf as resting spores (chlamydospores). It is spread in infected clippings or as spores on contaminated equipment. Hot, dry, windy weather is especially favorable for expression of the new disease because infected stem bases do not allow moisture to reach the upper portions of the plant. The disease occurs in areas that have been stressed for moisture and areas in full sun, and is also favored by excessive nitrogen fertilization.

Control. Irrigate frequently in summer and early fall to avoid moisture stress in the plants. Avoid heavy nitrogen applications. Use resistant varieties. No complete control with fungicides has been attained in California.

Fairy ring symptoms.

Fusarium blight on annual bluegrass (Poa annua): circular spots of diseased turf.

Fusarium blight on annual bluegrass (Poa annua): close-up of a diseased spot. Detail: crown rot stage

FUSARIUM PATCH

Symptoms. This disease appears as roughly circular, dead patches 1 to 2 inches in diameter, which may enlarge to 12 inches. Leaves first become water-soaked and then turn reddish brown, then bleached. Minute, white or pinkish, gelatinous spore masses occasionally are seen on dead leaves. Fungus threads, also white or pinkish, may be seen in early morning. The disease is common on annual bluegrass, *Poa annua,* and bluegrass. Ryegrass, bentgrasses, and fescues are also susceptible.

Cause. The fungus, *Fusarium nivale,* probably survives the hot, dry summer months as resistant hyphae in the turfgrass debris. Cold (32° to 60°F), moist conditions such as prolonged rainy periods in winter favor the disease, which usually appears first on shaded plants. The disease is common in northern and central California in the winter, but is rare in southern California.

Control. Reduce shade and improve soil aeration and water drainage. Avoid excess nitrogen fertilization, especially in the fall. Fungicides are more effective if applied late in the fall, before the disease develops.

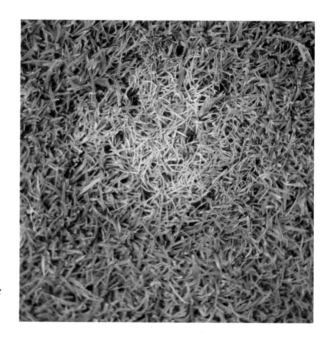

Fusarium patch damage to bentgrass (Agrostis spp.).

LEAF BLOTCH (BERMUDAGRASS)

Symptoms. Tiny, purplish to reddish spots occur on leaf blades and leaf sheaths. Seedlings are very susceptible, but growing plants rapidly become resistant. Affected seedlings wither, turn brown, and die. Roots and crown may develop small lesions. Only bermudagrasses are susceptible.

Cause. The fungus *Helminthosporium cynodontis* is similar to *H. vagans* and *H. sorokinianum*. It probably survives in infected bermudagrass plants and debris as hyphae and as spores, though it may be seedborne.

Control. Leaf blotch damages young seedlings or adult plants that have been weakened by excessive thatch, deficient nitrogen, and unfavorable conditions. Remove thatch and thin the turf at regular intervals. Apply adequate nitrogen.

LOOSE SMUT

Symptoms. In this disease of common bermudagrass, the flower heads are replaced by masses of dark spores.

Cause. The fungus, *Ustilago cynodontis*, remains in infected plants the year round. Spores infect germinating seeds and young stolons but show symptoms principally at flowering. The disease is most prevalent in warm weather and under conditions that promote flowering.

Control. Keep grass growing vigorously and remove flower heads by mowing before spores are produced. Because fungus spores may cling to seeds or occur in soil, seed treatment with suitable fungicides will prevent infection from seedborne or soilborne spores.

MELTING OUT (BLUEGRASSES)

Symptoms. Circular to elongated, purplish or brownish spots with straw-colored or brown centers occur on leaf blades, leaf sheaths, and stems. Leaf spots are general, indicating spread by windborne spores. Frequently, crown and roots are attacked. Crown-infected plants weaken, and many die during hot, dry, windy weather, resulting in a thinning out of the turf in scattered areas. Although common Kentucky bluegrass is very susceptible, several improved cultivars are resistant.

Cause. Melting out is caused by *Helminthosporium vagans*, which probably survives in infected plants or debris as fungus threads and as spores. It also may be seedborne. Cool (50° to 70°F), moist conditions favor the disease, which first appears on shaded plants and is most severe on closely clipped turf. Sometimes it is found on plants growing on compacted soils. This disease is common in the wet winter months, especially in coastal northern and central California.

Control. Reduce shade and improve soil aeration and water drainage. Avoid driving heavy equipment on soils. Do not mow bluegrass lower than 1½ inches. Several fungicides are helpful in controlling melting out if applied when the disease first appears. It is important to reduce the crown infections that are responsible for the death of the plants.

MELTING OUT (OTHER GRASSES)

Symptoms. Melting out of bentgrasses, fescues, ryegrasses, and bermudagrasses is caused by *Helminthosporium sorokinianum* (= *H. sativum*), which also attacks bluegrasses. The symptoms are similar to those of melting out caused by *H. vagans,* except leaf spots usually show brown rather than straw-colored centers, and borders of spots are purplish to dark brown.

Cause. High temperatures (70° to 90°F), dew, and high humidity favor the disease, which first appears on plants stressed for moisture and in areas of full sun. The disease causes more damage on closely clipped turf, especially in the spring and summer in southern California. Plants die of crown infection, usually during hot, dry, windy weather.

Control. Control is the same as for *H. vagans*. Mow grass to recommended heights. Do not allow plants to become stressed for moisture.

POWDERY MILDEW

Symptoms. Grey-white, cobwebby growth appears on upper leaf surface, at first in isolated patches, and spreads to give a grey-white appearance to leaves. In advanced stages, leaf blades may turn pale yellow. All turfgrasses are susceptible, but the disease is most severe on Kentucky bluegrass.

Cause. Powdery mildew is caused by *Erysiphe graminis*. It is most severe in shady areas with high humidity and poor air circulation and with air temperatures of about 65°F.

Control. Improve air circulation and reduce shading.

Melting out disease on Kentucky bluegrass (Poa pratensis).

*Powdery mildew on Kentucky bluegrass (Poa pratensis): **uninfected area contrasted to surrounding infected area.***

Close-up of Kentucky bluegrass (Poa pratensis) leaves infected with powdery mildew.

PYTHIUM BLIGHT (GREASE SPOT)

Symptoms. In advanced steps, the turf is killed in small, roughly circular spots (2 to 6 inches) that tend to run together. Blackened leaf blades wither rapidly and turn reddish brown. Leaf blades tend to lie flat, stick together, and appear greasy. Roots may be stunted and brown, especially at the tips. In mild stages, plants may be weakened and more susceptible to other diseases. All turfgrasses are subject to attack by the fungi that cause grease spot.

Cause. The fungi that cause grease spot are called "water molds" because the excessive free water essential for fungal development may weaken the plants, favoring severe disease development. Several species of *Pythium* may be involved, especially *P. aphanidermantum*. The causal fungi produce thick-walled sexual spores that survive for long periods in the soil. Because the fungus depends on excessive free moisture, grease spot usually appears in low spots that remain wet. It is also more common on heavy soils. *Pythium aphanidermantum* can spread very rapidly at high temperatures (80° to 100°F) and cause severe damage; other species of *Pythium* may cause damage during moderately warm weather.

Control. Reduce shading and improve soil aeration and water drainage. Water, when needed, to a depth of 4 to 6 inches. Several fungicides are available.

RED THREAD

Symptoms. Turf is usually affected in patches 2 to 15 inches in diameter, but the disease may appear generally over a large area. Pink webs of fungal threads bind leaves together and pink, gelatinous fungal crusts ¼ to ¾ inch long may be seen projecting from leaves. Plants usually are not killed. Bentgrasses, bluegrasses, fescues, and ryegrasses are susceptible.

Cause. Red thread is caused by *Laetisaria fuciformis,* which overseasons as pink or red, gelatinous crusts of fungal threads. Disease commonly occurs along the coast of northern and central California, but rarely in southern California. Red thread usually appears on plants deficient in nitrogen and during periods of prolonged, cool, wet weather.

Control. Apply adequate nitrogen and reduce shading. Fungicides, if applied at the earliest stages of the disease, will prevent further development.

Pythium grease spot on annual bluegrass (Poa annua): infected spots surrounded by healthy grass.

Close-up of turf damaged by pythium grease spot.

Red thread (close-up).

RHIZOCTONIA BLIGHT (BROWN PATCH)

Symptoms. Areas of turf affected with brown patch consist of irregular brown areas that may range from a few inches to many feet in diameter. The center of a spot may recover, resulting in a ring of diseased grass. Leaves and sheaths turn olive-green, wilt, become light brown, and then die. Stems, crowns, and roots may also be infected. In light attacks, roots and crown usually are not involved and plants recover. All lawn grasses are susceptible.

Cause. Brown patch is caused by the soil-inhabiting fungus *Rhizoctonia solani* and other *Rhizoctonia* species. The fungi are active as fine threads (hyphae) that survive in the soil or in and on the turf. Resting structures (sclerotia) composed of hard masses of fungus threads resist adverse conditions and are difficult to control with fungicides. The disease is favored by excess thatch and mat, high temperatures (75° to 90°F), high humidity (99 to 100%), and the soft, lush growth caused by excess nitrogen. The disease is most common in summer and in warm, inland areas. A cold (40° to 60°F), wet-weather form of the disease is more rare.

Control. Reduce shading and improve soil aeration and water drainage. Water, when needed, to a depth of 4 to 6 inches if possible. Check this with a soil probe, since excess turf debris may keep water from penetrating, as may an unfavorable soil type or too rapid an application of water. Avoid excessive nitrogen fertilization, since it encourages soft growth of the foliage. Fungicides are more effective as preventive applications, but will also stop the disease in progress.

RUST

Symptoms. Elongated, reddish brown to orange pustules containing spores appear on the stems, leaves, and leaf sheaths. Reddish brown to orange spores adhere to fingers when pustules are rubbed. Some bluegrass varieties are particularly susceptible, though ryegrass is commonly infected.

Cause. *Puccinia striiformis, P. graminis,* and *P. coronata* overseason in infected grasses. The airborne spores may be carried long distances. Cool to moderately warm, moist weather favors rust development. Condensed moisture, even dew, for 10 to 12 hours is sufficient to allow spores to infect plants.

Control. Keep plants growing rapidly by applying nitrogen fertilization and irrigating. Most of the infected portions of leaves are removed by weekly mowing, and the spores trapped inside the mowed leaves die out rapidly. Removal of grass clippings from the turf may help.

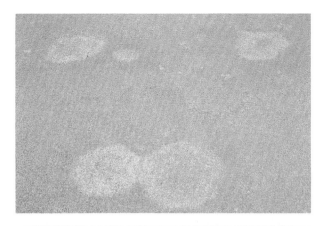

Rhizoctonia blight (brown patch).

Rust on bluegrass.

SEED ROT, DAMPING-OFF, AND ROOT ROT

Symptoms. Seed may rot in the soil or the young grass seedlings may be killed (damp-off) before or after emergence. Seed rot is not mushy but rather firm. In damping-off, seedlings at first are water-soaked; then they blacken, shrivel, and turn brown. Many affected seedlings are not killed but appear yellow and stunted, with markedly reduced root systems. All types of grasses are subject to seed rot and damping-off.

Cause. Seed rot, damping-off, and root rot are caused by several species of *Pythium,* by *Rhizoctonia solani,* and by *Helminthosporium* species. They are favored by excessive moisture, excessive nitrogen, seeds sown too deeply, and seeds of low viability sown above the recommended rates, especially during periods unfavorable for seed germination and growth.

Control. Improve soil aeration and water drainage. Do not overwater. Sow only fresh, healthy seed at recommended rates and seasons. Do not allow a soil crust to form. Additional fertilizer should not be applied until after the grass seedlings are established. Treat seed with one of several fungicides. Have a qualified expert fumigate the soil with a volatile fumigant before planting. Soil fumigation also kills weed seeds, insects, and nematodes.

SOUTHERN BLIGHT

Symptoms. Circular areas and rings of dead grass up to 9 feet in diameter occur in turf that usually is heavily thatched. These rings result from the outward growth of the causal fungus and the grass that recovers in the center (much like fairy ring). Inspecting the base of an affected grass plant within the ring while the fungus is active will reveal abundant white mycelial growth and small ($1/32$- to $1/16$-inch) spherical, light to dark brown sclerotia resembling mustard seeds. During late stages of the disease, only the brown sclerotia may be seen on the thatch. Because dead rings of grass can be caused by other fungi, the sclerotia are diagnostic.

Cause. Southern blight of turf is caused by the fungus *Sclerotium rolfsii.* In California, this fungus is known to attack ryegrasses, fescues, bentgrasses, and bluegrasses. *Sclerotium rolfsii* has a wide host range. It also attacks dichondra and many food crop plants. Formerly, it attacked only dichondra lawns, but, in recent years the fungus has been found attacking turfgrasses.

Control. Southern blight is becoming widespread in the central and southern regions of California and is favored by warm or hot weather, by moisture, and especially by thatch accumulation. The fungus is inactive in cold

Southern blight on mixed grasses: characteristic rings of dead grass.

Sclerotia of the causal fungus of southern blight, Sclerotium rolfsii.

weather. Because the sclerotia can act as "seeds" to spread and initiate the disease, take care to prevent their movement or the movement of infected plant parts. Effects of the disease may be reduced by good cultural practices. Thatch should be kept to a minimum by verticutting. However, aerating and verticutting equipment can spread sclerotia and infected plant parts.

SPRING DEAD SPOT

Symptoms. Circular areas of dead grass 6 to 12 inches in diameter occur as the turf resumes growth in the spring. Spots may coalesce to form large areas. The fungus survives in debris as hyphae and sclerotia, and is spread by sclerotia and infected plant parts. All bermudagrasses are susceptible but the disease is more severe on the hybrids.

Cause. Spring dead spot is caused by *Leptosphaeria korrae*. It is most severe during the winter when bermudagrass is dormant.

Control. Maintain healthy turf to speed recovery. Aerating and verticutting can spread the fungus.

STRIPE SMUT

Symptoms. Infected plants are often pale green and stunted, with long, black stripes consisting of fungal spore pustules in the leaves. Infected leaves curl, become shredded, and die. The disease is favored by moderate temperatures and is prevalent in the spring and fall. Infected plants may die in hot, dry weather. Although common on some bluegrass varieties, the disease also affects bentgrasses.

Cause. Spores of *Ustilago striiformis* from the leaves frequently contaminate seeds; germinating spores may infect seedlings and young tillers. The fungus stays in a diseased plant from year to year.

Control. Some resistant bluegrass varieties have been developed. Seed treatment with selected fungicides will control the seedborne phase. A systemic fungicide is available.

Spring dead spot on a home lawn. Detail: close-up of infected plant.

Stripe smut: infected bluegrass.

Stripe smut: infected bermudagrass.

TAKE-ALL PATCH

Symptoms. Circular or ring-shaped dead areas range from a few inches to 3 feet or more in diameter. Dying bentgrass at the advancing margins has a purplish tinge. The roots of diseased plants are rotted. Black strands of mycelium are visible on the surface of roots. Large black perithecia visible with a hand lens may occur on dead tissue. This is principally a disease of bentgrass.

Cause. The fungus *Gaeumannomyces graminis* var. *avenae* survives in grass debris and living grass plants. In California, take-all patch mainly occurs in the late fall and winter. Soil conditions that favor the disease include light texture, low organic matter, low or unbalanced fertility, and high pH. The disease is also favored by high moisture conditions.

Control. Improve growing conditions, drainage, pH, and fertility to limit the disease. Replant with less susceptible grasses.

Take-all patch surrounded by healthy turf.

7

Rodents and Related Vertebrate Pests

TERRELL SALMON, EXTENSION WILDLIFE SPECIALIST,
UNIVERSITY OF CALIFORNIA, DAVIS

In many situations, wildlife species are welcomed. However, they can be an economic threat in turf and landscape areas, and in these situations they must be controlled.

Rodent and related pests of turfgrass eat the above- and below-ground parts of grasses, plants, and trees. Their burrowing and mounding, particularly that of pocket gophers, disrupt maintenance operations, and in some cases actually cause damage to mowing equipment. Mounds also cover and kill nearby plants. Burrowing can damage structures, roads, and pathways as well. Not only is this damage expensive to repair, it may result in human injury (e.g., turning an ankle on a rodent burrow).

POPULATION CONTROL

When pest population control is necessary, develop an integrated approach based on knowledge of the animal's ecology and behavior as well as information on all available control techniques. This type of integrated pest management (IPM) program will result in an environmentally and economically acceptable approach that will significantly reduce damage to turfgrass.

Some people would institute control measures whenever pests occur. While this is appropriate for some wildlife pests in certain situations, a good IPM program is based on monitoring the pest to determine when control is necessary. When the population density reaches the threshold level—the level at which control is economically justified—control should be undertaken. Unfortunately, no threshold levels for rodent and other vertebrate pests in turf areas have been determined. Because of the nature of their damage to these areas, the tolerable level is very low, in some cases, zero.

The timing of a control program, as well as the methods and materials to use, depend on (1) the pest species, (2) how the area is managed, and (3) the availability of equipment and labor, as well as other factors. For specific information on pesticide solutions to problems caused by these animals, see UC Division of Agriculture and Natural Resources Leaflet 2209 (*Guide to Turfgrass Pest Control*).

Legal restraints. Ground squirrels, meadow voles, moles, and pocket gophers are classified as nongame mammals by the California Fish and Game Code. If you find them injuring or threatening your crops or other property, you may take them in any manner. You must satisfy special provisions of the California Fish and Game Code if you wish to use leg-hold, steel-jawed traps. These traps are not recommended for small rodents or rabbits.

Jackrabbits are game mammals according to the California Fish and Game Code. If they injure growing crops or other property, they may be taken at any time or in any manner by the owner or tenant of the premises. Cottontail (brush) rabbits are also game mammals, and may be taken by the owner or tenant of the land, or by any other person authorized in writing by such owner or tenant, when the rabbits are damaging crops or forage. Any person other than the owner or tenant of the land must be carrying written authority from the owner or tenant at the time rabbits are transported from the property. The rabbits cannot be sold.

The California Fish and Game Code lists skunks as nongame mammals and raccoons as furbearing mammals. Waterfowl are migratory game birds. Check with the local fish and game warden if you need to control any of these animals.

GROUND SQUIRRELS

The California ground squirrel (*Spermophilus beecheyi*) inhabits most agricultural and rural areas of California. Ground squirrels damage many food-bearing and ornamental plants. They damage young shrubs, vines, and trees by gnawing bark, girdling trunks, eating twigs and leaves, and burrowing around roots. Squirrels also gnaw surface-type plastic irrigation pipe.

Burrowing is probably the most destructive aspect of ground squirrel damage to turfgrass. In the process of digging burrows, ground squirrels heap up large mounds of soil and rock that may bury and kill grass or other small plants. Burrows make the turf difficult to mow and present hazards to machinery.

Ground squirrels frequently burrow around trees and shrubs, damaging the root systems and sometimes killing plants. Burrows beneath buildings and other structures can also cause damage.

Ground squirrels can transmit diseases such as tularemia and plague to humans, particularly when squirrel populations are dense. Do not handle dead squirrels. If you notice an unusual number of dead squirrels, notify public health officials.

Ground squirrels live in a wide variety of habitats, but populations may be particularly dense in areas disturbed by humans, such as road or ditch banks, fence rows, near buildings, and in or near many crops. They usually avoid thick chaparral, dense woods, and wet areas, and they live in colonies of 2 to 20 or more animals, spending much of their time in underground burrows.

Ground squirrels are active during the day and are easy to spot, especially in warm weather, from spring to fall. During winter, most ground squirrels hibernate, but some young squirrels remain active, especially where winters are not severe. Most adults go into a summer "hibernation" (estivation) during the hottest times of the year.

Ground squirrels reproduce once a year in early spring. Litter sizes vary, but seven to eight young are average. The young remain in the burrow for about 6 weeks before they emerge.

Ground squirrels are primarily vegetarians. During early spring, they consume green grasses and forbs. When the vegetation begins to dry, squirrels eat seeds, grains, and nuts.

CONTROL

When ground squirrels cause damage, you can institute a control program suitable for the situation and time of year. Figure 1 shows the activity cycle of ground squirrels, and indicates when control measures are appropriate. Such a program should result in significant reductions in your area's squirrel population.

Fumigation. You can kill ground squirrels in their burrows with several types of toxic gas, some of which require special permits from your County Agricultural Commissioner. Fumigation should not be used beneath buildings. It is most effective in the spring or at other times when soil moisture is high. The soil moisture contains the gas within the burrow system and prohibits its diffusion into the small cracks often present in dry soil.

Ground squirrels have large burrows that can have several entrances. Treat all entrances and then seal them. Re-treat any newly opened burrows. Fumigation is ineffective during periods of hibernation and estivation, because at those times squirrels plug their burrows with soil. The plug is not obvious to a person examining the burrow entrance.

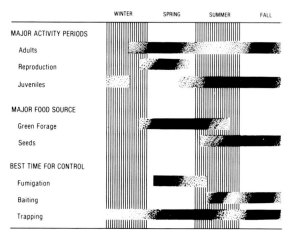

Fig. 1. *Many control methods are effective against ground squirrels only at certain times of the year. This chart shows the yearly activities of the California ground squirrel, and favorable times for baiting, trapping, and fumigating.*

Gases emitted from some fumigants occasionally ignite, creating a fire danger. Do not use such fumigants where a significant fire hazard exists—under buildings or near dry grass or other flammable materials.

Toxic baits (rodenticides). Some toxic baits are available over the counter, and others require a permit from your County Agricultural Commissioner. When you use toxic baits or any other rodent control materials, follow label instructions carefully.

Anticoagulant baits are recommended for control of ground squirrels because they are both effective against the pest and relatively safe to humans and pets. Anticoagulants interfere with an animal's blood-clotting mechanism, eventually leading to death. They are effective only when consumed in several feedings over a period of 5 or more days. Effectiveness is greatly reduced if 48 hours or more elapse between feedings. These features, as well as an effective antidote (vitamin K_1), make the use of anticoagulant baits relatively safe.

Anticoagulant baits can be used in either of two ways: in bait boxes or by repeated broadcast baiting. Bait boxes are small structures that the squirrel must enter in order to eat the bait (fig. 2). A box contains sufficient bait for repeated feedings, and helps keep children and pets from reaching the bait. The bait box is the preferred baiting method around homes and in other areas where children, pets, and poultry are present. Follow the product label regarding construction and placement of bait boxes.

There are several things you should consider when you are designing a bait box for ground squirrels. The entrance hole(s) should be about 4 inches across to allow access to squirrels but keep out larger animals. Construct a lip to prevent bait from spilling out of the box when squirrels exit. Provide a lock on the box or devise some other method that will make it difficult for children to open the box. The bait box should be secured so that it cannot be turned over or easily removed. A self-feeding arrangement will ensure that the pest gets a continuous supply of bait.

Place bait boxes in areas frequented by ground squirrels (near runways or burrows, for example). If ground squirrels are noticeable throughout the area, space the boxes at intervals of 100 to 200 feet. Initially, inspect bait stations daily and add bait if all is eaten overnight. Fresh bait is important. It may take several days before squirrels become accustomed to and enter the bait box. Anticoagulant bait generally takes 2 to 4 weeks or more to be effective. It does not immediately affect feeding habits of squirrels. Continue baiting until all feeding ceases and you observe no squirrels. Pick up unused bait and store it according to label instructions upon completion of the control program.

Repeated spot baiting (without a bait box) with anticoagulant bait can be effective in controlling ground squirrels. Follow label instructions. Generally, three or more treatments are necessary. If spot or broadcast baiting is not specified on the product label, do not use this baiting method.

Anticoagulant baits have the same effects on nearly all warm-blooded animals, including birds. Cereal baits are attractive to some dogs as well as to other nontarget animals, so take care to prevent their access to the bait. Danger to children and pets can be reduced by placing bait out of their reach in a bait box. Dead ground squirrels should be buried or put in plastic bags in the trash.

Single-feeding baits are also available for controlling ground squirrels. They are effective after only one feeding. Control is usually achieved in 1 to 3 days. Apply single-feeding baits by hand or with a mechanical broadcaster. Place the bait near the ground squirrel burrow or at places where the squirrels are feeding. Ground squirrels are good foragers and can easily find the broadcast grain. Do not pile bait, as this will increase the hazards to nontarget animals. As with all pesticides, follow label instructions carefully.

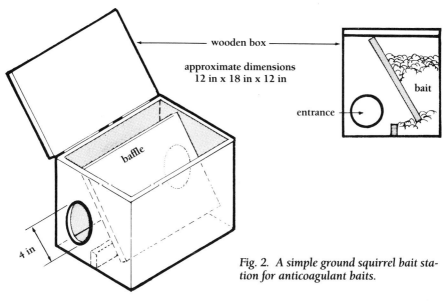

Fig. 2. A simple ground squirrel bait station for anticoagulant baits.

Trapping. Traps are practical devices for control of ground squirrels in small areas where the squirrels are few. Live-catch traps are effective but pose a problem: How to dispose of the animal? Because ground squirrels carry diseases and are agricultural pests, it is illegal to release them without a permit from the California Department of Fish and Game.

Several types of traps kill ground squirrels. Most work best if you place them on the ground near squirrel burrows or runways. Walnuts, almonds, oats, barley, and melon rinds are attractive trap baits. The bait should be well behind the trigger or tied to it. Bait the traps, but do not set them for several days so the squirrels will become accustomed to them. Rebait when the squirrels begin taking the bait, and this time set the traps.

A box-type squirrel trap (fig. 3) kills ground squirrels quickly. This trap is available commercially or can be constructed from a gopher box trap. To modify a gopher trap, lengthen the trigger slot with a rat-tail file or pocket knife so the trigger can swing unhindered and the squirrel can pass beneath the swinging loop of the unset trap. Remove the back of the trap and replace it with hardware cloth. This allows the animal to see the bait from both ends but prevents it from entering the trap from the back. For a dual-assembly trap, place two box traps back-to-back and secure them to a board (fig. 4).

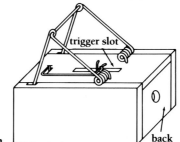

Fig. 3. Box-type squirrel trap.

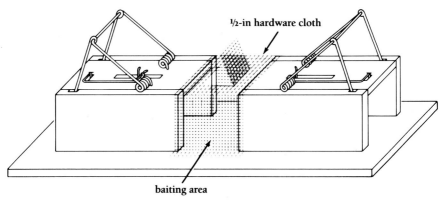

Fig. 4. Multiple-catch box-type ground squirrel trap.

Fig. 5. Conibear trap set in a burrow entrance. Secure the trap with a stake.

The Conibear trap is also an effective kill trap for ground squirrels. The wire trigger permits the trap to be used either with or, more commonly, without bait. Place the trap so the squirrel will pass through it, tripping the trigger. The best places to set the Conibear trap are directly in the burrow opening or where physical restrictions in the squirrel runway will direct the animal through the trap (fig. 5). Do not place it where pets or other nontarget animals are likely to pass. When you use a Conibear trap, leaving the trap baited but unset will have little effect on trapping success.

Other ground squirrel traps are available in some areas. With all traps, take precautions to reduce the hazard of trapping nontarget wildlife, pets, and poultry.

Natural control. As with all animals, natural constraints such as inadequate food and shelter, predators, disease, and bad weather limit ground squirrel populations. Experience has shown, however, that in most environments altered by humans, the point at which squirrel populations level off naturally is intolerably high.

Ground squirrels generally are found in open areas, although they usually need some cover in order to survive. Removing brushpiles and debris not only makes an area less desirable to ground squirrels, it also makes detection of squirrels and their burrows easier, aids in monitoring the population, and improves access during control operations.

Many predators—including hawks, eagles, rattlesnakes, gopher snakes, and coyotes—eat ground squirrels. In most cases, predators are unable to keep ground squirrel populations below the level at which they become pests. Predators sometimes can prevent ground squirrels from invading marginal habitats where cover is not abundant.

MONITORING GUIDELINES

Once you have controlled ground squirrel damage, establish a system to monitor the area for reinfestation. Observe from an isolated structure or automobile during the morning hours when squirrels are most active. Ground squirrels may move in from other areas and cause new damage within a short time. Experience has shown that it is easier, less expensive, and less time consuming to control a population before it builds up to the point where it can cause excessive damage.

MEADOW VOLES

Also known as meadow or field mice, meadow voles (*Microtus californicus*) damage a wide range of plants, feeding and gnawing on trunks, roots, stems, leaves, and seeds. They are common in rangelands and in many agricultural crops, and sometimes invade turfgrass.

Meadow voles are small rodents with heavy bodies, short legs and tails, and small, rounded ears. Their long, coarse fur is blackish brown to grayish brown. When fully grown, a meadow vole is 4 to 5 inches long.

Meadow voles are active all year and are normally found in areas dense with ground cover. They dig short, shallow burrows and make underground nests of grass, stems, and leaves. The peak breeding period comes in spring, with a second, shorter breeding period in fall. Litters average four young. Meadow vole numbers fluctuate from year to year; under favorable conditions, populations increase rapidly. Most problems around turfgrass occur during such times.

CONTROL

Preventing meadow vole damage usually requires a management program that restricts the population in the area. You can usually achieve this by removing or reducing the vegetative cover to make the area unsuitable for voles. By removing cover, you also make the voles easier to detect. These animals can cause severe damage and they have a rapid reproductive rate, so effective control depends on initiating a program of habitat modification or population reduction before their numbers explode around or on turfgrass.

Habitat modification. A particularly effective way to deter voles is to modify the habitat. Weeds, heavy mulch, and dense vegetative cover encourage meadow voles by providing food as well as protection from predators and other environmental stresses. If you remove the protective cover, the area will be much less hospitable for voles. By clearing dense grassy areas adjacent to turfgrass sites, you can help prevent damage and reduce the base area from which voles invade. Weed-free strips can also serve as buffers around protected areas. The wider the cleared strip, the less apt meadow voles will be to cross over and become established. A minimum strip width of 15 feet is recommended, but even that can be ineffective when vole numbers are high. Buffer strips are most useful around young trees or shrubs, or in other areas where the voles will have to remain in the open to feed.

Wire or metal barriers at least 12 inches high and with a mesh size of 0.25 inch or less will exclude meadow voles from turf areas (fig. 6). Meadow voles rarely climb such fences, but they may dig beneath them. To prevent digging, bury the bottom edge 6 to 10 inches deep.

Fig. 6. Small-mesh wire attached to the bottom of a fence will exclude meadow voles.

Toxic baits (rodenticides). When meadow voles are numerous or when damage occurs over large areas, you may need to use a toxic bait. Take care to ensure the safety of children, pets, and nontarget animals. Follow product label instructions carefully.

Anticoagulant baits are slow acting and must be consumed over a period of 5 or more days to be effective. They are, therefore, probably the safest type of rodent bait for use around developed areas. Many types and brands of anticoagulant baits are available.

Because the pest must feed on most anticoagulant baits over a period of 5 days, the bait must be available until the vole population is controlled. Bait placement is very important. Place it in runways or next to burrows so voles will find it during their normal travel (fig. 7). Usually, baiting every other day for 5 days will be sufficient. Be sure to broadcast the bait evenly over an infested area if that application method is specified on the label.

If so stated on the product label, paraffin bait blocks can be used in high-moisture areas. Place them in runways or near burrow openings, or both. Replace them as they are eaten and remove those that remain when feeding stops. Bait blocks should not be used where children or pets might find them or pick them up.

Baits that require only one feeding are called single-feeding baits. They are particularly useful where vole populations are spread over large areas. Place the bait in runways or next to burrows according to label instructions. A potential problem with these baits is "bait shyness," a condition that results when voles eat only enough bait to make them sick. If this happens, the voles will not eat the bait again for 6 months or more. To prevent this, use the bait according to label directions and do not treat with single-feeding baits more often than every 6 months.

Fig. 7. Place bait in a meadow vole runway or next to a burrow opening.

Fig. 8. Set snap trap perpendicular to meadow vole runway.

Single-feeding baits act quickly. You may find dead voles within 12 hours of baiting. Dispose of all dead voles by burying them or placing them in plastic bags and putting them in the trash. Do not handle the dead voles with your bare hands.

Trapping. Trapping is seldom used for control of meadow voles in turf areas. Sometimes, trapping can help you determine what species of animal is present, however (fig. 8).

Natural control. As with all animals, natural population constraints limit meadow vole numbers. Because populations will not increase indefinitely, one control alternative is to do nothing, letting the voles limit themselves. Experience has shown, however, that the natural population peak is often too high to prevent damage.

Predators, especially raptorial birds, eat meadow voles. However, in most cases, predators are unable to keep vole populations below damaging levels.

MONITORING GUIDELINES

To detect the presence of voles, look for fresh trails in the grass, as well as burrows, droppings, and evidence of feeding. Routine monitoring of the surrounding area is important. Pay particular attention to adjacent areas with heavy vegetation, because such areas encourage population buildups.

MOLES

Moles (*Scapanus* spp.) are small, insect-eating mammals, not rodents. In California, moles inhabit the Sierra Nevada and Coast Ranges and their foothills, as well as the entire coastal zone. They are not generally found in the Central Valley or in dry southeastern areas of the state. Moles live almost entirely underground in networks of interconnecting tunnels 3 to 30 inches deep. They feed mainly on worms, insects, and other invertebrates, but also eat some roots, bulbs, and seeds. Their burrowing sometimes dislodges plants and damages roots. Their mounds and ridges can be annoying in turfgrass areas.

Moles have cylindrical bodies with slender, pointed snouts and short, bare or sparsely-haired tails. Their limbs are short and spadelike. Their eyes are poorly developed and their ears are not visible. They have short, dense, velvety fur. A mole bears one litter of three or four young early in spring.

The mounds formed by moles are pushed up from an open center hole. The soil may be in chunks, and single mounds often appear in a line over the runway connecting them. Main runways are usually less than 2 inches in diameter and may be 16 to 18 inches below the surface. Surface-breeding burrows appear as ridges that the mole pushes up by forcing, not digging, its way through the soil just below ground level. Moles are active throughout the year, although surface activity slows during periods of extreme cold, heat, or drought.

CONTROL

Moles are difficult to control. Of the various techniques available, you may need to use a combination to control the pests.

Trapping. The most effective tool for mole control is the trap. Several different mole traps are available at hardware stores, nurseries, or direct from the factories.

By understanding the mole's behavior you can improve your trap set. When its sensitive snout encounters something strange in the burrow, the mole is likely to plug off that area and dig around or under the object. For this reason, a trap generally straddles or encircles the runway, or is suspended above it. Most mole traps operate on the theory that a mole will push its way through soil that blocks its tunnel as it would in the case of a natural cave-in.

The trap springs in response to pressure from the mole's body or the soil pushed against a triggering plate.

Moles are active throughout the year and can be trapped at any time. Before setting mole traps, determine which runways are in current use. Moles dig a system of deep tunnels as well as a network of surface runs. Many of the surface tunnels are used only temporarily in search of food. Trapping is most successful in the deeper tunnels.

To determine where moles are active, look for the freshest mounds. You can locate deeper tunnels by probing between, or next to, a fresh mole hill with a pointed stick, slender metal rod, or standard gopher probe (fig. 9). When the earth suddenly gives way to your probe, the probe has probably broken through the burrow. Another method is to stamp down short sections of surface runways. Observe these areas daily and restamp any raised sections, remembering the areas of activity. The selection of a frequently used surface runway is very important if you plan to set traps in it.

Several types of mole trap are available in California. The scissor-jaw type (fig. 10) and the harpoon type (fig. 11) are the most often used. Moles have

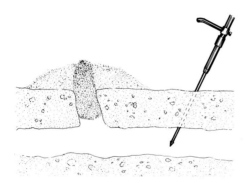

Fig. 9. Probe between or near mole mounds to find the main runway.

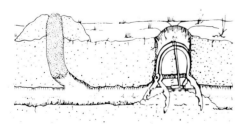

Fig. 10. Set the scissor-jaw trap so it straddles the runway. Remember to fill the portion of the tunnel under trap's trigger with loose soil.

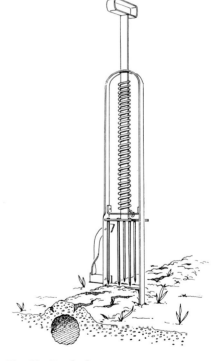

Fig. 11. Set the harpoon trap so its supporting stakes straddle the runway. This trap may also be set in the main runway like the scissor-jaw trap in figure 10.

also been caught with gopher traps set in mole runways. Most trap manufacturers provide detailed instructions that you should follow carefully.

Set the scissor-jaw trap in the mole's main underground tunnel, usually 6 to 10 inches deep. Using a garden trowel or small shovel, remove a section of soil slightly larger than the trap width (about 6 inches). Wedge the set trap, with safety catch in place, firmly around the opened burrow. Take care that the opened trap jaws do not protrude into the open ends of the mole's tunnel lest the animal become suspicious. Now sift loose dirt onto the set trap to about the level of the coil spring to exclude light from the opened burrow and make the mole less suspicious of the plugged tunnel. Release the safety catch, and the trap is set.

The harpoon trap will work in the deeper tunnels if you set it as described for the scissor-jaw trap, filling the hole with fine soil. It can also be set on the surface over an active runway ridge that has been pressed down under the trigger pan.

Toxic baits (rodenticides). Because the mole's main diet consists of earthworms and insects, poisoning with toxic baits is rarely effective, However, where a mole population covers a large area, poison baits may reduce the population to some degree. If you choose this technique, you will probably need to follow it with trapping.

Other methods. People have tried flooding or fumigating mole burrows with various gases and chemicals, but with little success. If moles are deprived of their food supply, they will move to other areas to find food. Therefore, an insect control program may indirectly result in reduced mole populations.

A number of other methods have been suggested to solve mole problems. These include placing irritating materials such as broken glass, razor blades, rose branches, bleach, mothballs, lye, and even human hair in the burrow, or using frightening devices such as mole wheels, vibrating windmills, and whistling bottles. Another reported method is the use of repelling plants, such as the gopher or mole plant (*Euphorbia lathyris*). None of these approaches has proven successful in stopping mole damage or in driving moles from an infested area.

MONITORING GUIDELINES

Once you have controlled damage, establish a system to monitor for reinfestation. Mounds and surface runways are easy to detect, and either one indicates reinfestation. Because mole damage is unsightly, the number of moles that can be tolerated in turfgrass is usually quite low, sometimes nil. As soon as you see an active mound or surface runway, initiate appropriate control.

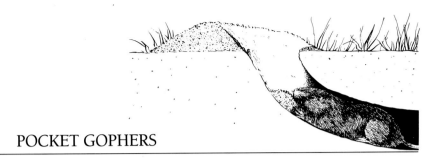

POCKET GOPHERS

Pocket gophers (*Thomomys* spp.) are stout-bodied, short-legged rodents, well adapted for burrowing. They live singly in extensive underground burrow systems that can cover several hundred square feet. These burrows are about 2 to 3 inches in diameter, and usually run from 6 to 12 inches deep. Pocket gophers often invade turf areas, ornamental plantings, and agricultural crops. They eat a wide variety of roots, bulbs, tubers, grasses, and seeds, and sometimes even the bark of trees. Their feeding and burrowing can damage lawns, ornamental plants, vegetables, forbs, vines, and trees. In addition, they may damage plastic water lines and lawn sprinkler sytems; their tunnels can divert and carry off irrigation water and lead to soil erosion.

A pocket gopher can range in length from 6 to 12 inches. It has a thick body with little evidence of a neck, and its eyes and ears are small. Its good sense of smell helps it to locate food. The common name is derived from the fur-lined external cheek pouches, or pockets, used to carry food and nesting materials. The pocket gopher's lips can be closed behind its four large incisor teeth, keeping dirt out of its mouth when it uses its teeth for digging.

Pocket gophers seldom travel above ground. You can sometimes see them feeding, pushing dirt out of their burrow systems, or moving to new areas. The mounds of fresh soil that result from burrow excavation indicate their presence. Such mounds are usually crescent-shaped and are located at the ends of short lateral tunnels branching from a main burrow system. One gopher may create several mounds in a day.

CONTROL

Because of the nature of pocket gopher damage, a successful control program depends on early detection and promptly applied measures appropriate to the location and situation. Most people control gophers in turfgrass areas with traps or poison baits. A program incorporating these methods should significantly reduce pocket gopher damage in your area.

Successful trapping or hand-baiting depends on finding the gopher's main burrow. The crescent-shaped mounds visible above ground are connected to the burrow by lateral tunnels. Because the gopher plugs its lateral tunnels, you will have little success trapping and baiting there.

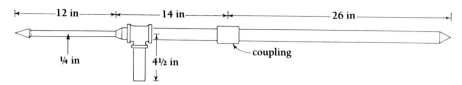

Fig. 12. *A probe for locating pocket gopher tunnels can be built at home. The shaft may be one piece or divided by a pipe coupling for convenient carrying.*

To locate the main burrow, use a gopher probe. You can purchase a gopher probe commercially or construct one from pipe, wooden dowel, or stick. An example is shown in figure 12. Look for the freshest mounds, since they indicate recent gopher activity. You will usually see a small circle or depression representing the plugged lateral tunnel. This plug is generally bordered on one side by soil, forming the mound into a crescent shape. Begin probing 8 to 12 inches from the plugged side of the mound. When the probe penetrates the gopher's burrow, it should drop suddenly about 2 inches. Often, the main burrow will lie between two mounds. To locate the gopher's main burrow, you will probably have to probe repeatedly, but your skill will improve with experience.

Trapping. You can control pocket gophers with traps. Several types and brands of gopher trap are available. The most common is a two-pronged pincher trap (fig. 13) that triggers when the gopher pushes against a flat vertical pan. Another popular version is the squeeze-type box trap.

After you have located the gopher's main tunnel, open it with a shovel or garden trowel and set traps in pairs facing opposite directions. This placement will intercept a gopher coming from either direction. The box trap is somewhat easier to set but requires more excavation because of its size. Box traps are useful when the diameter of the gopher's main burrow is small (less than 3 inches), because small burrows must be enlarged to accommodate wire traps. All traps should be wired to stakes so you won't lose track of them. After setting the traps, exclude light from the burrow by covering the opening with dirt, sod, cardboard, or some other material. Fine soil can be sifted around the edges to ensure a light-tight seal. If light enters, the gopher may plug the burrow with soil, filling the traps and making them ineffective. Check traps often and reset them when necessary. If no gopher is caught within 3 days, reset the traps in a different location.

Toxic baits (rodenticides). Several pocket gopher baits are available. Most are single-feeding baits, such as strychnine and are generally effective with one application, whether by hand or by mechanical applicator. Baits containing anticoagulants are also available for hand baiting. They require multiple treatments or one large treatment to be effective. All gopher bait is poisonous and should be used with caution. Read and follow product label instructions carefully.

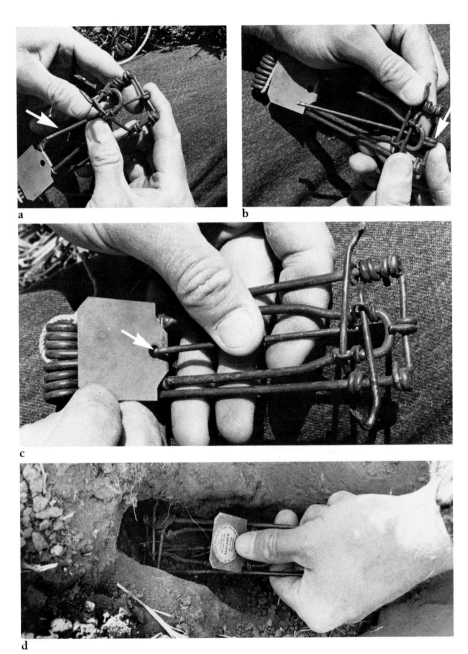

Fig. 13. To set the Macabee trap: (a) hold the trap exactly as shown in the illustration; be sure your index finger holds the trigger (arrow) up through the trap frame; (b) press your thumbs down to spread the sharp jaws; use your index finger to guide the trigger hook (arrow) over the end of the trap frame; (c) still holding the frame down, place the other end of the trigger into the small hole (arrow) in the plate; (d) place the set trap, sharp jaws first, into the gopher burrow.

Hand baiting. Always place pocket gopher bait in the underground tunnel. After locating the main gopher burrow with a probe (fig. 14), enlarge the opening by rotating the probe or inserting a larger rod or stick. Then place the bait carefully in the opening, taking care not to spill any on the ground surface (fig. 15). A funnel is useful for preventing spillage. Close the probe hole with sod, rock, or some other material to exclude light and prevent dirt from falling on the bait. Tamp down existing mounds so you can distinguish new activity. If mound building continues for more than 2 days after treatment with a single-feeding bait, or for more than 7 to 10 days with an anticoagulant bait, you will need to re-treat the burrow or use another control method.

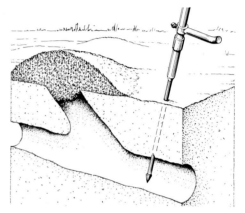

Fig. 14. Use your probe to place pocket gopher baits. When the probe suddenly drops about 2 inches, you have located a main tunnel. Enlarge the probe hole enough to insert poisoned bait.

Mechanical baiting. The mechanical bait applicator (fig. 16) offers an excellent way of controlling gophers over large areas with a once-over operation. This tractor-drawn device constructs an artificial underground burrow and deposits poison grain bait in it at preset intervals and quantities. The artificial burrow will intercept most of the natural gopher burrow systems. Gophers readily explore these artificial tunnels and will consume the bait. The percentage of active ingredient in the bait is usually higher for machine baiting than for hand baiting.

In turfgrass, the machine will cut the sod and leave a ridge. The machine can be useful for controlling gophers in areas surrounding turf and in rough areas where the ridge is not objectionable. With high gopher populations, the machine may do less damage than a large gopher population. Soil moisture is critical when you use the machine. If the soil is too wet, the tractor will bog down. If too dry, the artificial tunnel will cave in, resulting in poor control. In unirrigated areas, use the machine in late winter or early spring when natural moisture is high. Even in turf, soil is often too dry unless the area is heavily irrigated. For specifics on the machine and its use, consult University of California Division of Agriculture and Natural Resources Leaflet 2699 (*Pocket Gopher Control with Mechanical Bait Applicator*), available from your local farm advisor's office or from Agriculture and Natural Resources Publications.

Fig. 15. To probe and hand-bait for pocket gophers: (a) use your probe to find the gopher burrow; (b) after you feel a noticeable give, use the probe shaft to enlarge the hole; (c) insert a funnel into the hole and slowly pour bait down the funnel into the burrow; (d) remove the funnel and place a clod of dirt over the hole to block any light from entering the tunnel.

Fig. 16. The mechanical bait applicator mounts on a tractor's three-point hitch. When lowered, the hollow shank simultaneously forms a burrow and deposits poisoned bait.

Other methods. Pocket gophers can easily withstand normal irrigation, but flooding sometimes forces them out of their burrows where they become vulnerable to predation (usually human). Fumigation with smoke or gas cartridges is usually ineffective since gophers quickly seal off their burrows when they detect smoke or gas. Other fumigants may be effective, but little or no published information is available on their use in California.

No repellents currently available will successfully protect turfgrass from pocket gophers. The gopher plant (*Euphorbia lathyris*) has been suggested as a repellent, but there is no conclusive evidence to prove its effectiveness. Attempts to frighten gophers with sounds, vibrations, or other means are ineffective. Predators, especially owls, eat pocket gophers, but in most cases they are unable to keep pocket gopher populations below levels that cause problems in turfgrass areas.

MONITORING GUIDELINES

Once pocket gopher damage has been controlled, establish a system to monitor the area for gopher reinfestation. Level all existing mounds after the control program and clean away weeds and debris so fresh mounds can easily be seen. A monitoring program is important because pocket gophers can move in from other areas, bringing new damage within a short time. Experience has shown that gophers are easier, less expensive, and less time consuming to control before they build up to populations that can do excessive damage.

RABBITS

Rabbits can be destructive in turf and landscaped areas, especially when their populations are high. Rabbits eat a wide variety of plants, including grasses, grains, alfalfa, vegetables, fruit trees, vines, and many ornamentals. They also damage plastic irrigation systems.

Three rabbit species are common to California: the jackrabbit (*Lepus californicus*), the cottontail (*Sylvilagus audubonii*), and the brush rabbit (*S. bachmani*). Because it is larger and more abundant, the jackrabbit is the most destructive of the three.

A jackrabbit is about as large as a house cat. It has long ears, short front legs, and long hind legs. Jackrabbits typically occupy open or semi-open lands in California valleys and foothills. They do not build nests but make depressions in the soil beneath bushes or other vegetation. When born, young jackrabbits are fully haired and their eyes are open. Within a few days, they can move about quite rapidly.

Cottontail and brush rabbits are smaller and have shorter ears. They generally inhabit places with dense cover, such as brushy areas, wooded areas with some underbrush, or areas with piles of rocks or debris. Their young are born naked and blind, and remain in the nest for several weeks.

CONTROL

A number of methods can be used to reduce rabbit damage. Physical exclusion, trapping, and poison baits are recommended for protecting turfgrass.

Exclusion. If properly built, fences can be very effective in keeping rabbits out of an area. A 30- to 36-inch-high wire fence of a mesh no larger than 1 inch, the bottom of which is turned outward and buried 6 inches in the ground, will exclude rabbits (fig. 17). The fence must include tight-fitting gates with sills to keep rabbits from digging below the bottom rails. Keep gates closed as much as possible, since rabbits can be active day or night. Inspect the fence regularly to make sure rabbits or other animals have not dug under. Poultry netting supported by light stakes is adequate for rabbit control, but larger animals (especially livestock) can easily damage it. Cottontail and brush rabbits will not jump a 2-foot fence. Jackrabbits ordinarily will not jump a

Fig. 17. Rabbit fences must be attached securely into the ground. Burying the wire base will help prevent rabbits from digging under.

fence this high unless chased by dogs or otherwise frightened. Discourage jumping by increasing the aboveground height to 3 feet. Remember, once a rabbit gets into the fenced area it may not be able to get out.

Toxic baits (rodenticides). Both single- and multiple-feeding poison grain baits are available for jackrabbit control. Bait acceptance is often difficult, so take care in developing and conducting the baiting program.

Place the bait in areas frequented by rabbits (near trails and resting and feeding areas). Place bait according to label instructions for the particular material selected. Single-feeding baits require prebaiting with unpoisoned bait. This is extremely important because it conditions the rabbits to eat the bait and also to feed regularly at the baiting site.

Trapping. Trapping is ineffective for jackrabbits because of their reluctance to enter a trap. Trapping with a box or similar trap can be effective for cottontail and brush rabbits if their numbers are not great. Another simple way to trap rabbits is to construct a small corral along a rabbit-tight fence surrounding the protected area. Construct a short strip of fence at a diagonal to the main fence, funneling the rabbits through a one-way gate into the corral. Inspect the corral daily. Because rabbits can carry certain diseases and are considered agricultural pests, it is illegal to release them in other areas.

Other methods. Guns or dogs can be effective in eliminating small numbers of rabbits. Best results are achieved in the early morning or in the evening, when rabbits are most active. Check local regulations for any restrictions on shooting in your area.

Natural control. Predators, especially hawks and coyotes, eat rabbits. However, in most cases, these predators are unable to keep rabbit populations below damaging levels.

MONITORING GUIDELINES

Rabbits are large and easily seen, but because they frequently feed in darkness, you may have to examine the feeding area at night with a flashlight. Additionally, look for such signs as droppings, trails, and feeding damage. Generally, if rabbits are feeding in an area, droppings can be found nearby. Rabbits observed nearby are likely to invade the turfgrass area when the plantings become desirable to them. Therefore, consider exclusion methods or possibly an area-wide control program before damage actually occurs.

OTHER RELATED PESTS

Other vertebrate pests can and do cause problems on turfgrass areas. Skunks and raccoons often feed on grubs and other insects in turf. While feeding, they dig up the sod in search of food. Though no chemical repellents or toxicants are registered to eliminate or reduce this problem, several approaches may be helpful. Exclusion with fences can be effective, but is impractical for large areas. Insect control to reduce the food supply in turf can sometimes prompt skunks or raccoons to move on to areas where food is more plentiful. Another option is to contact governmental or private contractors who control these animals.

Coots, geese, and other waterfowl graze on turf and can cause significant damage and inconvenience. As with skunks and raccoons, no repellents or toxic baits are registered to reduce or eliminate the problems caused by these birds. Your local California Department of Fish and Game Office, County Agricultural Commissioner, or farm advisor should be able to direct you to agencies or individuals who specialize in controlling these pests.

Glossary

A. E. Acid equivalent; the expression for the active ingredient in a liquid herbicide formulation.

Aeration. The act of supplying or impregnating soil with air.

A. I. Active ingredient; the killing component in a pesticide formulation.

Annual. A plant that completes its life cycle from seed in one year.

Bait box. A small structure in which anticoagulant bait is placed. The target animal must enter to take the bait.

Biennial. A plant that completes its life cycle from seed in two years. The first year, it produces leaves and stores food; the second year, it produces flowers and seeds.

Broadcast application. Application of a material over an entire area.

Broadleaf weed. The popular term for a plant in the dicotyledon (dicot) group; e.g., dandelion, knotweed, or plantain.

Broad-spectrum biocide. A pesticide that is effective against many insects or microorganisms.

Burrow. An underground excavation used by some rodent and related pests for shelter, food storage, and rearing of young.

Chlorophyll. The green photosynthetic coloring matter of plants found in chloroplasts and made up chiefly of a blue-black ester and a dark green ester.

Compatability. The quality showed by two or more chemicals or products that can be mixed without affecting each other's performance.

Contact herbicide. A herbicide that kills only the plant tissue contacted, with little or no translocation.

Contact insecticide. An insecticide that kills its intended victim when the victim touches or is touched by it.

Control. Regulation or restraint of the population of a species that has become a pest.

Culm. The erect stem of a grass.

Cultivar. A crop of a kind originating and persistent under cultivation.

Dichondra. Any species of a genus of chiefly tropical perennial herbs related to the morningglory, including some plants used for ground cover and as a substitute for lawn grasses in warmer parts of the United States.

Dicotyledonae (dicot). The botanical subclass in which dicotyledon (broadleaf) plants are grouped.

Emulsion. A suspension of one liquid in another, as distinguished from a *solution,* in which the two liquids combine completely and become one.

Estivation (aestivation). A state of dormancy in an animal's activity cycle that occurs during the summer.

Exclusion. The prevention of one or more animals' or species' entry to an area.

Flower stalk. The stem bearing a flower on a plant.

Foliage-applied herbicide. A herbicide whose primary action is on or through the foliage.

Forb. Any small, broadleaf flowering plant.

Formulation. A manufactured blend of a pesticide and other ingredients available as a liquid, wettable powder, or granule for pest control.

Fumigant. A substance that produces toxic or suffocating gases.

Fumigation. The use of a chemical that is put into the soil as a gas or in a form that will turn into gas to destroy unwanted plants (weeds), insects, nematodes, or disease-causing elements.

Game mammal. A mammal, as specified by the California Fish and Game Code, to be hunted for food or sport.

Habitat modification. The alteration of the environment where an animal species is found. In wildlife pest control, this sometimes consists of making the habitat less favorable for the pest species.

Herbicide. A chemical used for killing or inhibiting the growth of plants.

Hibernation. A state of dormancy in an animal's activity cycle that occurs during the winter.

Inert ingredient. An inactive component of the pesticide formulation.

Internode. The part of a stem between two successive nodes.

Leg-hold trap. A device for trapping and holding or restraining an animal by the leg.

Metrics (S.I.). An international system of measurement based on the meter, liter, and gram.

Midrib. The middle or main vein of a leaf.

Monocotyledonae (monocot). The botanical subclass in which monocotyledon (narrowleaf) plants are grouped.

Multiple-feeding bait. A poisonous bait that requires a sustained dosage over a period of time to produce death; e.g., an anticoagulant.

Mycelia. The filamentous vegetative structures of fungi.

Narrowleaf weed. The popular term for a plant in the monocotyledon (monocot) group; e.g., all grasses, nutsedge.

Natural control. The regulation or restraint of an animal population in a manner consistent with nature; e.g., by predators, diseases, or lack of food or cover.

Node. A joint, as in a stem; the point where buds and leaves occur.

Nongame mammal. Any mammal not commonly hunted, as specified by the California Fish and Game Code.

Nonselective herbicide. A weed control chemical that kills plants in general, without regard to species.

Nontarget species. Any species that is not the object of the pest control application.

Obligate. An organism that is restricted to one particularly characteristic mode of life.

Ornamental. A plant that is cultivated for its aesthetic value.

Panicle. The many-branched flower head, with flowers at the end of each branch, found in many grasses; e.g., annual bluegrass.

Perennial. A plant that lives for more than 2 years.

Photosynthesis. The formation of carbohydrates in the chlorophyll-containing tissues of plants exposed to light.

Phytotoxicity. The quality of being injurious and sometimes lethal to plants.

Postemergence. The period after a plant's germination and emergence from the soil.

Prebaiting. The practice of placing nontoxic bait at a feeding site to condition the pest to eat it before you apply toxic bait, or of placing bait in an unset trap so the pest will become accustomed to the trap before it is set.

Predator. Any animal that survives by regularly killing other animals for food.

Preemergence. The period before a plant's germination and emergence from the soil.

Preplant. Any time before a crop is planted.

Prostrate. The general term for plants tending to grow flat on the ground.

Rhizome. An underground stem with nodes and internodes, usually producing roots and stems or leaves from the nodes. *See* Stolon.

Rootstock. In turfgrass, a term often used interchangeably with rhizome.

Rosette. A cluster of closely crowded leaves arising from a very short stem near the surface of the ground.

Runway. The path commonly traveled by an animal or animals.

Sclerotia. The compact mass of hardened mycelium stored with reserve food material that in some higher fungi becomes detached and remains dormant until a favorable opportunity for growth occurs.

Seedling. An infant plant grown from seed.

Selective herbicide. A weed control chemical that will kill some plant species when applied to a mixed population, without causing serious injury to other species. Excessive herbicide application rates may reduce or eliminate the desired selectivity.

Shoot. A young, aboveground branch or growth.

Single-feeding bait. A toxic bait that produces death from one dose. Also called *acute toxic bait.*

Sod. A large section of turf. The term is usually used in the context of turf production for vegetative propagation.

Soil-applied herbicide. A chemical that is taken up by plants principally from the soil.

Soil fumigant. A material used to control a broad range of organisms, usually a vapor or gas that diffuses through the soil and has a relatively short life in the soil. *See* fumigation.

Spike. An inflorescence (flower head) with spikelets (flowers) attached directly to an unbranched stem.

Spikelet. One of a small group of grass flowers composing a spike.

Spot baiting. Placing bait by hand at selected sites.

Spot spraying.. The application of a herbicide concentration to individual weeds, applying no more spray than is required to wet the foliage.

Stolon. A stem that grows horizontally along the ground, usually rooting at the nodes. A new plant may be formed from each section of stolon that roots. *See* Rhizome.

Summer annual. A plant that germinates in the spring, grows to maturity during the summer, develops seeds, and dies during the fall or winter.

Surfactant. A material that improves the emulsifying, dispersing, spreading, wetting, or other surface-modifying properties of an herbicide formulation. Its action is similar to that of a detergent.

Systemic. A pesticide that is absorbed into and distributed throughout a plant.

Taproot. A stout, vertical root giving off small lateral roots.

Thatch. The layer of undecomposed or partially decomposed stems and leaves at the soil surface.

Translocation. The movement of a herbicide within a plant from the point of entry to other areas; e.g., from the leaves to the roots.

Untreated bait. Bait that is free of toxic or repellent substances; used for prebaiting.

Vascular system. The transport avenues in plant tissue, such as the leaf veins.

Vein. A conductor or transport unit of a plant's vascular system.

Verticutting. The act of cutting with blades that move perpendicularly to the soil surface. The purposes are to thin turf, control grain, and aid in control and elimination of thatch.

Viability. The quality of being alive and capable of germinating and growing; generally applied to seed.

Weed control. The process of limiting a weed infestation so that desired plants can be grown.

Weed eradication. The complete elimination of all live plants, plant parts, and seeds from an area.

Wetting agent. A compound that, when added to a spray solution, causes it to contact plant surfaces more thoroughly. *See* Surfactant.

Wildlife pest. Any species of wild animal, in any area, that becomes a health hazard, causes economic damage, or is a general nuisance to one or more persons.

Winter annual. A plant that germinates in the fall or winter, lives through the winter in the vegetative stage (often a low rosette), flowers and seeds in the spring, and dies in the summer.

WARNING ON THE USE OF CHEMICALS

Pesticides are poisonous. Always read and carefully follow all precautions and safety recommendations given on the container label. Store all chemicals in their original labeled containers in a locked cabinet or shed, away from food or feeds, and out of the reach of children, unauthorized persons, pets, and livestock.

Recommendations are based on the best information currently available, and treatments based on them should not leave residues exceeding the tolerance established for any particular chemical. Confine chemicals to the area being treated. THE GROWER IS LEGALLY RESPONSIBLE for residues on the grower's crops as well as for problems caused by drift from the grower's property to other properties or crops.

Consult your County Agricultural Commissioner for correct methods of disposing of leftover spray material and empty containers. **Never burn pesticide containers.**

PHYTOTOXICITY: Certain chemicals may cause plant injury if used at the wrong stage of plant development or when temperatures are too high. Injury may also result from excessive amounts or the wrong formulation or from mixing incompatible materials. Inert ingredients, such as wetters, spreaders, emulsifiers, diluents, and solvents, can cause plant injury. Since formulations are often changed by manufacturers, it is possible that plant injury may occur, even though no injury was noted in previous seasons.

The University of California, in compliance with Titles VI and VII of the Civil Rights Act of 1964, Title IX of the Education Amendments of 1972, Sections 503 and 504 of the Rehabilitation Act of 1973, and the Age Discrimination Act of 1975, does not discriminate on the basis of race, religion, color, national origin, sex, mental or physical handicap, or age in any of its programs or activities, or with respect to any of its employment policies, practices, or procedures. Nor does the University of California discriminate on the basis of ancestry, sexual orientation, marital status, citizenship, medical condition (as defined in Section 12926 of the California Government Code) or because individuals are special disabled veterans or Vietnam era veterans (as defined by the Vietnam Era Veterans Readjustment Act of 1974 and Section 12940 of the California Government Code). Inquiries regarding this policy may be addressed to the Affirmative Action Director, University of California, Agriculture and Natural Resources, 300 Lakeside Drive, 6th Floor, Oakland, CA 94612-3560. (415) 987-0097.

5m–rev–2/89–WJC/FB